About Island Press

Since 1984, the nonprofit organization Island Press has been stimulating, shaping, and communicating ideas that are essential for solving environmental problems worldwide. With more than 1,000 titles in print and some 30 new releases each year, we are the nation's leading publisher on environmental issues. We identify innovative thinkers and emerging trends in the environmental field. We work with world-renowned experts and authors to develop cross-disciplinary solutions to environmental challenges.

Island Press designs and executes educational campaigns in conjunction with our authors to communicate their critical messages in print, in person, and online using the latest technologies, innovative programs, and the media. Our goal is to reach targeted audiences—scientists, policymakers, environmental advocates, urban planners, the media, and concerned citizens—with information that can be used to create the framework for long-term ecological health and human well-being.

Island Press gratefully acknowledges major support of our work by The Agua Fund, The Andrew W. Mellon Foundation, The Bobolink Foundation, The Curtis and Edith Munson Foundation, Forrest C. and Frances H. Lattner Foundation, The JPB Foundation, The Kresge Foundation, The Oram Foundation, Inc., The Overbrook Foundation, The S.D. Bechtel, Jr. Foundation, The Summit Charitable Foundation, Inc., and many other generous supporters.

The opinions expressed in this book are those of the author(s) and do not necessarily reflect the views of our supporters.

RESILIENT BY DESIGN

RESILIENT BY DESIGN

Creating Businesses That Adapt and Flourish in a Changing World

Joseph Fiksel

ISLANDPRESS

Washington | Covelo | London

Island Press is a trademark of The Center for Resource Economics.

Library of Congress Control Number: 2015948278
ISBN: 978-1-61091-587-8 (cloth)

Printed on recycled, acid-free paper ✪

Manufactured in the United States of America
10 9 8 7 6 5 4 3 2 1

Keywords: Adaptability, business processes, crisis, design for resilience, innovation, management and decision-making systems, market complexity, natural resources, organizational design, organizational resilience, resilience, risk management, stability, supply chains, sustainability, turbulence

Contents

Foreword

Resilient by Design by Joseph Fiksel is an important book on the most important subject of our time. For companies and organizations of all kinds that are navigating the rapids of accelerating technological, social, and economic change, mastery of the art and science of resilience will be the difference between thriving, surviving, and extinction. The word *resilience* implies a combination of flexibility, adaptability, and foresight; it is a close kin to the word *sustainability* for the simple reason that no organization could long survive without the capacity to anticipate and accommodate change.

As Fiksel explains, the core principles of resilience are straightforward but become more difficult to manifest as enterprises age and become victims of prior success. These principles are, I think, virtually the same for other kinds of organizations, including institutions of higher education and governments. In all cases, resilience implies a change in outlook, perspective, and thinking by which enterprises learn to thrive in turbulence, thus becoming what Nassim Nicholas Taleb calls "antifragile." Further, in all organizations, resilience requires learning to see the enterprise and the world beyond as patterns and systems that are often unpredictable. Accordingly, resilient organizations leave wide margins of safety to accommodate surprise and the unknown. They do not bet the house on a single roll of the dice.

Fiksel is a superb guide to the art and science required to navigate the rapids ahead, but even in the commercial world of "creative destruction," some things must not change. Legitimate enterprises of all kinds must deliver what they promise at a fair price, a concept otherwise called integrity. They must compete, certainly, but they must also learn to cooperate so as to protect the health and stability of the system of which they are a part. They must improve quality and service while lowering ecological and social costs. From these steps one can discern the outlines of a wider dialogue about resilience.

Do the products and services of more resilient enterprises and organizations contribute to the resilience of the broader economy? If they do not, should such products be made or such services rendered regardless of market demand? If they should not be made or rendered, how does the enterprise or organization grow into a better and more constructive line of work? What does resilience possibly mean on our progressively hotter and more threadbare Earth? What does resilience mean on a planet losing the biological diversity and ecological stability essential to resilience itself? What is the future of resilient enterprises on this ecologically brittle Earth?

The ultimate test of resilient enterprise is not whether it is flexible and creative enough to survive and beat the competition for a while longer as things fall apart around it. The truest test of resilience is whether the transition fosters wider circles of resilience that cascade into a world that becomes more just, decent, and durable, which is to say life-centered. That would be a world that preserves things of enduring value, including children's lives, wildness, and beauty. In other words, the transition to resilience, enterprise by enterprise and organization by organization, is not an end in itself but rather a means to preserve and enhance the enterprise of life itself.

David Orr
Counselor to the President
Oberlin College

Preface

As a firm believer in environmental and social responsibility, I spent most of my management consulting career advising companies on how they could adopt sustainability strategies, technologies, and business processes. I built the sustainability practice at Battelle, a prominent technology firm, and became vice president for life cycle management. By 2000, though, I had become increasingly impatient over the slow pace of change. While industry leaders were embracing sustainability goals and reporting their accomplishments, it was evident that their fundamental business models had not changed. What's more, global economic growth, especially in developing nations, was outstripping any well-intentioned efforts to slow down our consumption of energy, water, soil, and other natural resources. Even today, despite widespread concerns about manifestations of climate change, efforts to mount a serious response are met with ambivalence and political opposition.

In 2002, a revelation set me down a new path. First, I realized that no company could be expected to compromise its essential mission of creating value for shareholders. As long as the current business model appeared to be working, the sustainability program would remain an appeasement tactic, simply expanding regulatory compliance to include compliance with stakeholder expectations. Initiatives that saved money, such as waste recovery, were acceptable, but there was not much appetite for transformative change. Second, I recognized that most company executives were consumed with the day-to-day burdens of managing a complex enterprise and had little patience for pondering hypothetical scenarios about the future. The primary purpose of the enterprise was to survive crises, adapt to change, and continue to flourish. Then it dawned on me that this was exactly the purpose of living systems, from the tiniest microbes to the mightiest nations. Their driving force is not the hope for sustainability, it is the necessity for resilience.

As I began to investigate this concept, I discovered that resilience had already been observed and studied in many different fields, such as anthropology, psychology, medicine, biology, evolution, ecology, engineering, and management. It appeared, however, that no one had tried to bridge these disciplines and unify knowledge about resilience, that the linkage between resilience and sustainability had never been adequately understood, and—perhaps most surprising—that no one had tried to apply the lessons of resilience in living systems to the challenges of enterprise management. It was a white space waiting to be explored. In 2003, I published a seminal paper, "Designing Resilient, Sustainable Systems," and launched on a journey of discovery.

Working with my esteemed colleague Bhavik Bakshi, I developed financial support to establish a new research center at The Ohio State University called the Center for Resilience. Our mission was to improve the resilience of industrial systems and the environments in which they operate, with the premise that short-term risk management and long-term sustainability are two ends of the enterprise resilience continuum. Thus began the most creative, productive, and satisfying period of my life. We assembled an advisory board of prominent companies, worked on a variety of innovative grants and contracts, issued a number of influential publications, and joined a community of practice around the world that was beginning to assemble the pieces of the puzzle. About a decade later, resilience seems to be emerging as a new strategic imperative. It took some major technological failures, political upheavals, an economic recession, and several natural disasters, but world leaders are finally understanding that we need to become more resilient at both the national and local levels. Ironically, climate change is now seen as one of the greatest perils that we face. It's not hypothetical any more.

As companies grapple with the challenges of the hyperconnected twenty-first-century economy, they are beginning to pay attention to the resilience of their critical assets: people, property, resources, and reputation. The risks of conducting business as usual are no longer acceptable, and traditional methods for managing those risks are no longer effective. Perhaps the most daunting challenge is the complexity and interdependence of environmental, social, and economic systems, making it difficult to assess the hidden consequences of innovative technologies and business practices. Companies will need to expand their planning boundaries to consider the infrastructure, the built environment, the ecosystems, and the social fabric in which they and their business partners operate.

Resilient by Design is intended as a guide to executives and managers who are taking on the task of building a more resilient enterprise. The most powerful lever for enhancing resilience is design, in the broadest possible sense. The scope of design must expand from products and processes to the enterprise as a whole, exploring how changing external conditions might influence business success. Our recommended approach to "design for resilience" considers the health and viability of important external systems, including stakeholders, communities, infrastructure, supply chains, and natural resources. Thus, design will become less of a rigid specification exercise and more of a dynamic intervention in ongoing cycles of change. To understand these complexities, businesses will begin to collaborate more closely with government, academia, and nonprofit groups. Best of all, this strategy may be turn out to be a practical path toward the elusive long-term goal of sustainability.

I want to acknowledge some of the outstanding individuals who have helped me over the years. David Miller of Island Press was instrumental in shaping and perfecting this book. My valued colleagues and key supporters at Ohio State have included Bud Baeslack, Bhavik Bakshi, Kate Bartter, Keely Croxton, Aparna Dial, Casey Hoy, Elena Irwin, Richard Moore, Marc Posner, Rajiv Ramnath, Phil Smith, Kathy Sullivan, Dave Williams, and Dave Woods. As an advisor to the US Environmental Protection Agency, I had the privilege of working with Derry Allen, Paul Anastas, Karen Chu, Gary Foley, Herb Fredrickson, Michael Gonzalez, Iris Goodman, Alan Hecht, Lek Kadeli, John Leazer, Montira Pongsiri, Subhas Sikdar, Cindy Sonich-Mullin, Marilyn ten Brink, Barb Walton, and other talented people too numerous to mention. I am thankful to the many others who have helped me on this journey, including Andrea Bassi, Emrah Cimren, Bob Costanza, Peter Evans, Peter Fox-Penner, Tom Hellman, Mike Long, Andy Mangan, Oleg Mishchenko, David Orr, Tim Pettit, Jed Shilling, Kieran Sikdar, Jerry Tinianow, the late Warren Wolf, and Darrell Zavitz. Most importantly, I thank my wife, Diane, for her sage advice and constant encouragement, making me a happier and more resilient person.

Resilience teaches us that we can't achieve a utopian steady state because it is not realistic and perhaps not even desirable. We live in a world of perpetual change, including cycles of growth and collapse. Some companies will prosper, and others will decline. There will be catastrophes and reversals of fortunes. There will be renewal of old industries and growth of new industries based on new ideas. In this increasingly dynamic world, we need to ensure that the things we care about deeply are resilient and able to survive the inevitable turbulence.

We may seem complacent, but we can be ingenious and powerful when our comfortable existence is threatened. Now is the time for resilience.

PART 1

Resilience as Competitive Strategy

CHAPTER ONE

Embracing Change

*The greatest danger in times of turbulence is not the
turbulence—it is to act with yesterday's logic.*

Peter Drucker[1]

Today's interconnected, global economy is characterized by turbulence. Markets are volatile, supply chains are increasingly vulnerable, and disruptions can substantially affect shareholder value. Major disasters, be they natural or caused by humans, can occur unexpectedly. Even minor incidents such as a local power failure can cause significant financial losses. Emerging pressures such as climate change and urbanization will only intensify the potential for extreme events and business interruptions. At the same time, these shifting conditions are opening up new market opportunities.

The word *turbulence* suggests a river of change, constantly in motion, with many waves and eddies both large and small, slow and fast. That largely describes today's business landscape. Steering an enterprise through this turbulent environment has become an exercise in alertness and rapid adaptation, akin to white-water rafting, and the waves of change are coming faster and harder. It's enough to keep any company executive awake at night.

What are the options for companies to cope with turbulent change?

- Resist change by hardening defenses and trying to maintain stability.

- Anticipate change by preparing for disruptions based on experience and foresight.

- Embrace change by designing an organization that can adapt to unforeseen challenges.

The premise of this book is that to succeed in the face of turbulence, enterprise managers will need to anticipate and embrace change rather than resist it. The problem is that we still tend to cling to a belief in stability as the normal state of affairs. When a disaster strikes, such as a hurricane or a terrorist attack, our instinct is to overcome the shock, assist the victims, and return to a stable equilibrium as soon as possible. But what if the quest for stability is futile? Faced with a turbulent business environment, our best strategy may be to plunge in, accept change as the new normal, and improve our capacity for rapid response and adaptation. To ride the waves of change, companies need to become more resilient. They need to be prepared for unexpected events and bounce back quickly or, better yet, "bounce forward" by improving their competitive posture.

Turbulence is a consequence of many shifting forces, including cultural, political, technological, and environmental changes. These forces can be divided into two major types:

1. **Gradual stresses** include population growth, climate change, urbanization, mobile device proliferation, and the rising income gaps between the poor and the wealthy. Some types of gradual change, such as metal corrosion or sea-level rise, may not be recognized until severe consequences become evident.

2. **Sudden shocks** include hurricanes, tsunamis, industrial accidents, power failures, economic collapses, terrorist attacks, and political upheavals. In some cases, a small-scale disruption, such as a facility structural failure or a regulatory policy change, can trigger a chain of events that develops into a crisis.

Any of these forces alone would be challenging to cope with, but when they occur simultaneously and interact with one another, the challenges can seem overwhelming. A potent example occurred in 2013, when Superstorm Sandy pounded the northeastern coastline of the United States, which has gradually become more vulnerable to flooding due to rising sea level. As a result of this storm, much of the New York coast and New Jersey lost power and water service for weeks, and economic losses totaled about $70 billion. Our traditional management tools, such as risk analysis, are inadequate for understanding or predicting the collective effect of these complex forces on a business enterprise. Catastrophic disruptions that arise from an interplay of stresses and shocks are difficult or impossible to forecast with any confidence.

What Is Resilience?

Resilience is the capacity to survive, adapt, and flourish in the face of turbulent change.[2]

The most common use of the term is in human psychology. A resilient person is able to recover from adversity, such as a traumatic accident or a job loss, and forge ahead with confidence.

At a broader scale, resilience can be seen in social and cultural organizations, such as tribal, ethnic, or religious groups, as well as entire cities and nations.

Resilience is intrinsic in all living things. For example, bacteria are able to develop resistance to antibiotics. Likewise, ecosystems can recover from extreme damage such as an oil spill.

The resilience of a business enterprise is more complex because it depends on the resilience of people, products, processes, assets, markets, and communities.

Experience has shown that business enterprises tend to lose their resilience as they grow and mature. They become vulnerable to surprises and slow to recover from disruptions. Companies that emphasize stability may cling to outmoded practices and proven technologies, may fail to question their assumptions, and may have blind spots that hamper their recognition of external change. As a consequence, they are unable to react to external challenges until they reach a state of crisis and require a drastic intervention.

On the flip side, companies that embrace change are better positioned to identify and seize emerging opportunities more nimbly than their competitors. Today, innovative companies such as Dow Chemical, IBM, Unilever, and Royal Dutch Shell have begun to view resilience as a source of competitive advantage. They are supplementing their traditional risk management processes with continuous monitoring of external situations and strategic capabilities for agility and adaptation. Like skilled athletes, these companies strive to operate at peak performance while being alert and prepared for emerging challenges. As a consequence, they are able to thrive in a constantly changing environment, discerning opportunities and consistently building shareholder value.

Despite the turbulence around them, resilient companies find a way to survive and prosper. They accept the inevitability of surprises and are able to adapt gracefully, sometimes transforming their very structure. In the words of Andrew Grove, former chief executive officer (CEO) of Intel, "Bad companies are destroyed by crises; good companies survive them; great companies are improved by them."[3]

The New Normal

Crises are becoming more commonplace than ever. The giant reinsurance company, Munich Re, reported that there has been a sharp increase in the number of natural catastrophes since 1980, a trend that has been linked to climate change.[4] Other destabilizing pressures include rapid urbanization, resource depletion, and political conflicts. As our planet's systems become more tightly coupled and volatile, the incidence of "black swan" events seems to be increasing.[5] Aside from natural disasters, we are increasingly confronted with unexpected technological failures, including infrastructure collapses, power failures, and ecological crises such as BP's Deepwater Horizon oil spill of 2010 in the Gulf of Mexico.

Perhaps the greatest stress factor is the increasing complexity and connectivity of the networked global economy. Companies can no longer operate as isolated entities that focus on internal process improvement; rather, they must account for interdependencies, partnerships, and potential conflicts with suppliers and customers throughout their spheres of operation. For large multinational companies, this practice effectively covers the entire world. Thanks to the growth of international trade, industrial parts and feedstocks are sourced from distant parts of the world, and the resulting products are often exported to distant markets. As a result, companies may be vulnerable to shocks or stresses that are far from their view and generally outside their control.

For example, on March 11, 2011, a magnitude 9 undersea earthquake off the coast of Japan caused a powerful tsunami that swept away homes, businesses, and entire cities, claiming more than fifteen thousand lives. In addition, the earthquake and tsunami severely damaged the Fukushima Daiichi nuclear power station, triggering the greatest nuclear crisis since the Chernobyl events of 1986. Millions of households in Japan lost power for months, and radioactive contamination will remain a concern for years. Moreover, the ripple effects of this catastrophic event were felt by businesses around the world. The prolonged shutdown of many Japanese manufacturing plants created costly delays in part shipments for electronics, motor vehicles, and other industries. All told, the direct costs of the disaster were more than $200 billion, not even counting the worldwide losses due to business interruption. Besides natural disasters, there are many other types of shocks that can interrupt the continuity of global supply chains. A particularly worrisome issue in the United States is the threat of catastrophic failures due to increasing demands on aging infrastructure.

Another important stress factor is the increasing resource footprint of the globalized economy. We have become dependent on a massive global throughput of resources, including minerals, fuels, food, and manufactured goods. It has been estimated that the average US citizen accounts for movement of about 30 tons of material per year, with most being released as waste and emissions within a short space of time.[6] This excessive material consumption will only increase with rapid economic growth and increasing affluence in developing nations. It represents a clear threat to the sustainability of the world economy. The Global Footprint Network estimates that, if current trends continue, by the 2030s we will need the equivalent of two Earths to support the world's population.[7] To flourish, companies must ensure the resilience of the critical ecological resources that are vital for continued economic prosperity.

Finally, the information technology revolution has ushered in a new era of instantaneous communication, virtually unlimited computing power, and access to enormous volumes of data. The business implications for enterprise innovation and transformation are enormous and are beyond the scope of this book. From a resilience perspective, these developments represent a double-edged sword. On the one hand, they enable real-time situational awareness and more rapid response to unexpected events. On the other hand, the growing interconnectedness of people, organizations, buildings, vehicles, and electronic devices within what is called the Internet of Things only exacerbates the complexity of the overall systems that we attempt to manage.

Indeed, modern communication has created new vulnerabilities, as illustrated by comparing the 1989 *Exxon Valdez* and 2010 Deepwater Horizon oil spills. The *Exxon Valdez* incident was mainly a concern to local stakeholders, and Exxon was able to take a slow, deliberate approach to its cleanup operations and legal defense tactics. In contrast, the Deepwater Horizon incident was shared instantaneously via broadcast and social media, including real-time video of crude oil spewing into the Gulf of Mexico. This coverage placed BP at an immediate disadvantage in defending its actions and negotiating a settlement, resulting in enterprise-wide reputational damage.

Designing for Resilience

We used to think of a company as an efficient, well-oiled machine, but machines can break down when a crisis occurs. In fact, mechanistic systems based on strict logical rules cannot cope with events that their designers failed to anticipate. Engineered systems, including electronic devices, buildings, and utility networks, are vulnerable to sudden failure or collapse. They are generally *brittle*, just the opposite of resilient. Technological advances such as artificial intelligence can help improve robustness, but engineering solutions tend to focus on known challenges rather than prepare for the unexpected.

In contrast, resilient companies are able to avoid crashing because they behave like living organisms, sensing, responding, and adapting to change. In the natural world, resilience is seen everywhere from individual cells to entire ecosystems. Similarly, human beings possess extraordinary resilience at many different scales, from individuals to cities to entire cultures. It turns out that companies have a unique advantage. Rather than letting natural selection take its course, they can quickly adapt to a changing environment by redesigning themselves!

This book raises some simple questions: How can our management and decision-making systems operate more like living things and less like brittle machines? How can we better cope with unforeseen disruptions that threaten business continuity and profitability? How can we design our products, processes, and assets to be inherently resilient? How can doing so help us gain competitive advantage?

Embracing change and building inherent resilience will require a new approach to dealing with risk and uncertainty. The objectives of traditional processes such as enterprise risk management and business continuity management are to minimize unwanted disruptions and to quickly resume normal operations. These approaches are suitable for a stable environment with predictable changes that occur intermittently and independently. In today's more complex risk landscape, however, these approaches are inadequate for dealing with fast-moving, unfamiliar changes that may cascade into disasters. The most damaging disruptions are often a result of rare events that seem highly unlikely until they actually happen; the catastrophe at the Fukushima Daiichi nuclear power station is but one of many examples. Resilience implies the capacity to overcome changes that are not predictable or quantifiable, representing unforeseen threats and opportunities. In the absence of predictive information, resilience involves capabilities for sensing of discontinuities, rapid adaptation, and flexible recovery or transformation.

At the same time, new business opportunities are emerging in every field to help individuals, communities, and companies adapt to an environment of rapid change and increasing variability. As illustrated in table 1.1,

Table 1.1. Examples of adaptation to variability

	Home Heating System	Public Transit System
Eliminate variability	Home thermostat set to constant temperature	Adherence to a strict departure schedule
Manage variability	Thermostat adapts to time of day, home occupancy, and/or outside temperature	Mobile devices receive real-time information about schedule changes or delays
Embrace variability	Multiple residential heating and cooling technologies deployed by a "smart" controller	On-demand, mixed-mode transportation options with variable real-time pricing

such opportunities can range from eliminating variability (which is often futile) to managing variability to actually incorporating variability into products, processes, and services (often the most effective approach). An extreme example of embracing variability is the concept of mass customization, wherein every customer receives a unique product tailored to his or her specific needs.

Finally, in a tightly connected world, resilience is important not only to individual companies but to the global economy as a whole. In fact, resilience is a first step toward achieving the long-term goals of global sustainability. Responding to the pervasive challenges of water scarcity, climate change, and poverty, companies like IBM and Shell are working with communities to help them become "smarter" by redesigning urban management practices and infrastructures to improve quality of life and ensure continuity in the event of disasters (see chapter 11). Besides generating new markets, these companies are learning how to strengthen the resilience of critical ecosystem services such as flood regulation and soil formation, which provide the life support system for their global supply chains.

Strategies for Enterprise Resilience

In recent years, a growing number of multinational enterprises have launched efforts to improve the inherent resilience of their global operations. They have found that the lessons of resilience are applicable to every enterprise activity, from strategic planning to product development to operations management. They are better able to respond to disruptive forces and better able to seize business opportunities that may open up. Case studies of companies that demonstrate such practices appear throughout this book and are titled "Resilience in Action."

The term *resilience* has quickly entered the corporate lexicon, but there are as many definitions of the word as there are business functions. For example, some management theorists define strategic resilience as "the ability to dynamically reinvent business models and strategies as circumstances change."[8] Others prefer to define resilience in operational terms as an extension of business continuity management, as in "the ability to recover from unexpected disruptions" including chemical spills, information technology failures, natural disasters, or terrorist attacks.[9] In the broadest sense, enterprise resilience encompasses many familiar concepts, such as agility, adaptability, robustness, and continuity, but it goes beyond these tactical notions to the very heart of the enterprise structure and culture. Our definition of enterprise resilience is quite simple: "Resilience is the capacity of an enterprise to survive, adapt, and flourish in the face of turbulent change and uncertainty."[10]

From this perspective, resilience is not just the ability to bounce back quickly and recover from a disruption. Rather, resilience is a strategic approach to embracing change that addresses both downside and upside possibilities. Resilient enterprises continue to grow and evolve to meet the needs and expectations of their shareholders and stakeholders. They adapt successfully to turbulence by anticipating disruptive changes, recognizing new business opportunities, building strong relationships, and designing resilient assets, products, and processes.

Resilience is not a substitute for the established methods of enterprise risk management; rather, it enables companies to embrace change in a turbulent and complex business environment by expanding their portfolio of capabilities. Early adopters of resilience have demonstrated how they can augment traditional risk management practices with new competencies that help them anticipate, prepare for, adapt, and recover from disruptions and, in some cases, treat disasters as an opportunity for gaining competitive advantage. Companies like General Electric, IBM, and Swiss Re see the emerging interest in resilience as an opportunity for new products, services, and markets.[11]

With the increasing pace and unpredictability of change, resilient companies have shifted from a reactive mode to the adoption of proactive and adaptive strategies, and accordingly have implemented a variety of strategic responses. As depicted in figure 1.1, these responses can be divided into four categories, depending on the magnitude and abruptness of changes that occur both inside and outside a company.

1. **Steer and adjust:** When the pace of change is slow and manageable, involving relatively minor fluctuations, companies can use orderly, well-defined business processes that operate precisely and efficiently. The concept of continuous improvement, based on a "plan-do-check-act" cycle, enables periodic midcourse corrections to ensure that companies learn from experience and achieve ever-higher performance

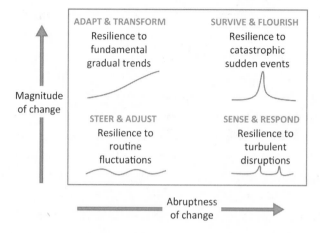

Figure 1.1. Embracing change through resilience strategies

goals. An example of the steer and adjust strategy is inventory management based on seasonal demand forecasting.

2. **Sense and respond:** Every business may experience unexpected disruptions that interfere with normal business operations. Disruptions can range from known risks, such as fires and chemical spills, to black swan events that are difficult to anticipate. Risk analysis and emergency response procedures help anticipate common types of disruptions and ensure business continuity. For disruptions that are rare or unpredictable, companies can supplement traditional risk management processes with the capacity to sense early warning signals and respond in a flexible manner.

3. **Adapt and transform:** Gradual changes in the business environment may eventually erode a company's competitive advantages. By using trend forecasting and scenario planning, companies have become more proactive in identifying major paradigm shifts that could influence their strategy, such as the growth of Internet commerce and the emergence of new market segments. To adapt to such trends, many companies have turned to reengineering and change management, although internal change is often difficult and true business transformation is rare.

4. **Survive and flourish:** Increasing globalization and complexity have amplified the turbulence of the business environment, forcing

companies to abandon reactive approaches and become wary of future predictions. Catastrophic disruptions are becoming more common, and it has become clear that "business as usual" is a fallacy. Disaster recovery is merely a survival strategy. To remain successful and flourish under these challenging conditions, companies must anticipate possible futures, develop adaptive capacity, and embed resilience thinking into their business processes.

Ironically, the original architects of the modern business enterprise designed for stability, essentially resisting the relentless waves of change rather than moving with them. The quality movement, which sought to eliminate product and process variability, appears to be a successful strategy up to a point. Establishing precise schedules, standardizing work processes, and emphasizing repeatability have resulted in greater operating efficiencies, higher yields, and associated cost savings. Today, increased variability in the business environment requires greater flexibility in business processes, implying a shift from rigid, prescriptive processes to more fluid processes that are sensitive to changing conditions. Likewise, the "lean" movement, which sought to eliminate waste from business processes, has made production systems more susceptible to unplanned disruptions. Optimization can actually weaken resilience by removing buffers that protect against fluctuations. Instead, practitioners now advocate a middle ground— lean and agile—that balances waste elimination with the need for flexibility and backup resources.

Embracing change and embedding resilience requires not only a continued focus on internal process excellence, but also an awareness of emerging patterns in externally coupled systems, including regulatory, socioeconomic, and environmental changes. One highly visible example is Business for Innovative Climate and Energy Policy (BICEP), an advocacy coalition of more than two hundred US companies working with policy makers to pass energy and climate legislation that will enable a rapid transition to a low-carbon economy. BICEP companies have recognized that, regardless of scientific uncertainties, climate change concerns cannot be ignored by major producers or consumers of energy. They are positioning themselves for competitive advantage in the face of anticipated changes such as carbon taxes and emission limits.

Table 1.2. Initiatives for improving enterprise resilience

	Functional Processes	**Structural Configurations**
Steer and adjust	Tracking and forecasting	Supply chain flexibility
	Performance monitoring	Real-time communication
	Continuous improvement	
Sense and respond	Early warning systems	Back-up and reserve capacity
	Emergency preparedness	Transportation alternatives
	Risk management protocols	Alliances and partnerships
	Business continuity plans	
Adapt and transform	Trend analysis and learning	Diversity of capabilities
	Business process redesign	Organizational agility
	Change management	Business model innovation
Survive and flourish	Scenario-based planning	Asset security and fortification
	Crisis management	Modularity and redundancy
	Opportunity realization	Business diversification

Table 1.2 gives an overview of various initiatives that contribute to enterprise resilience and is divided into both functional and structural considerations for each of the above strategies. The time scales for these initiatives can vary from weeks to years. Functional initiatives range from increased agility in recognizing and resolving problems (e.g., emergency preparedness) to fundamental transformations in response to strategic threats or opportunities (e.g., business process redesign). Structural initiatives range from establishing safeguards against disruptions (e.g., supply chain flexibility) to reducing vulnerability to change and increasing versatility (e.g., business diversification).

Insights into the resilience of complex systems can help in the design of more resilient business models as described in chapter 10. For example, a collection of distributed electric generators (e.g., fuel cells) connected to a power grid can be more resilient than a central power station in handling disruptions. That is a form of structural resilience. Similarly, a geographically dispersed workforce linked by telecommunications may be less vulnerable to catastrophic events that could disable a centralized facility. Research has shown that a dynamic enterprise fluctuating between different organizational structures (e.g., centralized vs. decentralized) can be more efficient than one that adheres to a constant model. Indeed, as conditions changed over several decades, major companies such as Hewlett-Packard and Ford have deliberately oscillated between a centralized and decentralized structure.[12]

Using This Book

The rest of *Resilient by Design* explores these resilience strategies in greater detail and provides in-depth case studies of companies that are acknowledged leaders in enterprise resilience. Achieving the overarching vision of a resilient enterprise that embraces change will require embedding inherent resilience into corporate assets and business processes. It is not a simple proposition, but we have little choice. In a world of increasing turbulence and complexity, resilience is essential for long-term economic prosperity and sustainability, both for corporations and the societies in which they operate.

The book is divided into three parts:

- The balance of part 1 explains why traditional risk management methods must be augmented by efforts to improve resilience and adaptability. We have much to learn from living systems, which are inherently resilient and constantly adapting to change. Nature does not require strategic planning.

- Part 2 turns to the practical questions of how resilience generates shareholder value and how it can be incorporated into business processes, including supply chain management, environmental management, and human resource management. A variety of tools and indicators are available to assess and manage resilience.

- Part 3 offers guidelines for taking a systems approach to "design for resilience" that considers important linkages among the enterprise, external communities, and ecosystems. The final chapter describes how short-term resilience can provide a path toward long-term sustainability, provided that we understand the trade-offs.

This book suggests that enterprise resilience is not just a strategy or a skill, but a fundamental attribute that can be designed into a company's assets, including technology, capital, people, organizational structure, products, and processes. Some enterprises have improved their resilience by shifting from a traditional, mechanistic style of management to a more organic, adaptive approach. In addition, leading companies are looking beyond their own boundaries and enhancing the resilience of the economic, ecological, and social systems to which they are coupled. Resilience provides a new language for enterprises to understand and embrace turbulent change, simultaneously addressing business prosperity and the broader needs of society.

Resilience on the Gridiron

Football competition is very well structured and less complex than the business world. Learning about resilience patterns in football could yield valuable insights for business enterprises. Examples of resilience in football include coping with setbacks such as loss of a key player, changing the offensive strategy to overcome a competitive challenge, and maintaining confidence during a losing streak. Resilience can be seen at different time scales: play by play, series by series, game by game, and even season by season.

To test this idea, I spoke with Archie Griffin, the legendary two-time winner of the Heisman Trophy who has served as president of the Ohio State Alumni Association. Griffin views resilience as an integral part of a successful football program. He identified a number of important aspects of resilience:

- **Perseverance.** If you are knocked down, you bounce back up. This principle was instilled into Archie by his parents at an early age.

- **Preparation.** Practice for every eventuality you can imagine. Archie's fabled coach at Ohio State, Woody Hayes, used to rehearse pass plays constantly even though his team was known as a running team.

- **Adaptability.** In the words of Hayes, "For every action there is a reaction and a re-reaction." Football is like a chess match, with teams constantly adapting to their opponents' tactics or to unexpected setbacks. Unwillingness to change is a recipe for failure, so you need to take chances and get out of your comfort zone.

- **Learning.** There are always lessons to be learned from defeat. Archie recalls Ohio State's 1973 contest against archrival Michigan in which his team's running game encountered a stiff defense. Ohio State did not attempt a pass and could only salvage a 10–10 tie. In contrast, during the next game against the University of Southern California, the team mixed pass and run effectively, and scored a resounding 42–21 victory.

- **Responsibility.** Mutual trust and accountability are important for team success. When every player feels responsible for supporting his or her teammates, the player strives to avoid letting the team down. Coach Hayes used to say, "In a team situation, you are better than you think you are."

- **Culture.** Perhaps most important of all is the team's belief in its power to overcome adversity. The greatest coaches are able to establish a culture of winning, and a defeat only spurs the team to work harder the next time.

These principles translate readily into the business world. In Archie's words, "Football is just a game, but business is for real. And football is great practice for real life."

According to Archie, there is no better example of resilience than Ohio State's 2014–2015 football season. Star quarterback Braxton Miller was injured in the preseason and was replaced by a talented but untested freshman, J. T. Barrett. After an early loss to Virginia Tech, Ohio State's dreams of a national championship seemed hopeless, but Barrett and his teammates displayed remarkable tenacity and did not lose another game during the regular season. Then adversity struck again. Late in the season, Barrett was injured, and third-string quarterback Cardell Jones was thrust into the limelight. But Coach Urban Meyer had prepared his team: Ohio State overwhelmed Wisconsin 59-0 in the Big Ten championship and proceeded to an improbable postseason run, defeating mighty Alabama and Oregon to claim the national football championship.

Takeaway Points

- In the new normal, crises are becoming more commonplace, and the ripple effects on business enterprises can be felt around the world.

- Rather than resisting change and clinging to stability, innovative companies are developing adaptive capacity to overcome adversity and seize opportunity.

- Enterprise resilience is defined as the capacity to survive, adapt, and flourish in the face of turbulent change and uncertainty.

- Depending on the magnitude and abruptness of change, a variety of resilience strategies are needed, including both functional and structural approaches.

CHAPTER TWO

From Risk to Resilience

*Our world is dominated by the extreme, the unknown, and
the very improbable . . . while we spend our time engaged in
small talk, focusing on the known and the repeated.*

Nassim Nicholas Taleb[1]

Embracing change means accepting uncertainty about the future.
Of course, making business decisions is challenging when the outcomes are uncertain. An entire academic field called management science
offers mathematical tools to help companies make all sorts of decisions,
from setting inventory targets to making major capital investments. Central to this field is the concept of *risk*, a term that is widely used and often
misunderstood.

Upside Downside

There are two main interpretations of risk, depending on whether you
take an operational perspective or a financial perspective. In the world of
operations, a risk is viewed as a threat to business continuity, that is, the
possibility of an adverse event resulting in loss, damage, or injury. From
this perspective, risk is a *downside* to be avoided and ideally eliminated; it
is a defect in an otherwise smoothly operating system.

In contrast, the financial world views risk as a fundamental measure
of variability. Every business activity or investment of resources involves
some downside risk, but we pursue these activities and investments with
the hope of gaining an *upside* reward. Typically, the larger the risk, the
larger the potential reward. Savvy managers and investors have learned to

discern the important business characteristics that increase the likelihood of success.

Both types of risk can be represented mathematically by a "risk profile," a curve that gives the likelihood of positive or negative outcomes. These outcomes can be measured in monetary terms (e.g., dollars gained or lost) or in a variety of other ways (e.g., market share gained or lost). Generally, the procedure of describing risks in this way is called risk analysis or risk assessment, and it can entail a lot of effort to gather the necessary information. Without good historical information, it is especially difficult is to estimate the probabilities of various outcomes.

Figure 2.1 illustrates a typical risk profile for a hypothetical situation, such as a capital investment.[2] Although there is a high likelihood of a positive return (upside), there is also a considerable chance of a loss. The best case and worst case are at the extreme ends of the curve, but in reality it is difficult to pinpoint a maximum gain or maximum loss. In fact, as shown in figure 2.1, we often underestimate the possibility of rare events that could result in catastrophic losses, as indicated by the "fat tail" of the curve. Statisticians have found that complex, interconnected systems often follow a *power law* pattern; for example, an event of magnitude x might occur with a likelihood of $1/x^2$. In other words,

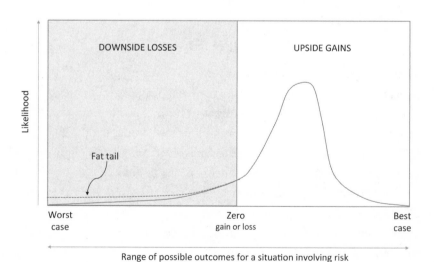

Figure 2.1. Example of a risk profile

extreme events are much more likely to happen than is predicted by the commonly used *normal*, bell-shaped distribution, which assumes independence among system components.[3] The power law explains the apparent frequency of extreme disruptions, such as hurricanes, stock market swings, and traffic jams.

If the future resembles the past, we can often construct risk profiles with a fair degree of confidence based on historical data. For example, in the property and casualty insurance industry, actuarial tables provide a reliable basis for setting premiums, but problems can arise if shifting conditions make the historical observations irrelevant. Moreover, the low-probability ranges, both upside and downside, represent rare outcomes that may never be observed in practice. For these reasons, development of risk profiles usually requires a large amount of estimation and subjective judgment, including the use of modeling and simulation. In many cases, risk assessment becomes a subjective exercise based on the beliefs of experts or decision makers, so pessimistic and optimistic assumptions may differ widely.

Clearly, risk management is a gloomy business if we focus only on downside risks. In this case, it is natural to strive for risk minimization and cling to stability. A more positive approach is to recognize that risk is inherent in competition and growth and thus view every setback as an occasion for learning and adaptation. Interestingly, the Chinese word for "crisis" consists of two characters that signify danger and opportunity (see figure 2.2). Global companies recognize that they must take calculated risks to grow and prosper in a business environment fraught with uncertainty.

Figure 2.2. *The Chinese characters for "crisis"*

Enterprise Risk Management

During the 1990s, the upside and downside views of risk were consolidated under a practice known as enterprise risk management (ERM), which has become the prevailing approach in large corporations.[4] ERM provides an integrated strategic process for identifying the portfolio of risks that are faced by various businesses, determining the corporation's "risk appetite" for each line of business, and using risk control strategies, including insurance, to achieve the appropriate level of risk. Most companies appoint a chief risk officer to oversee the ongoing implementation of ERM. Likewise, for the US Environmental Protection Agency (EPA) and other agencies, risk management has become the cornerstone of regulatory decision making.[5]

Corporate attention to risk management was heightened by several highly visible events, such as the mass deaths in Bhopal, India, caused by an inadvertent release of poisonous gas from a Union Carbide plant in 1984. Further motivation was provided by public entities, including the introduction of International Organization for Standardization (ISO) standards, Security and Exchange Commission requirements for management disclosure of "material" risks, and regulations such as Germany's Control and Transparency in Entities Law.[6] Concerns about catastrophic risks have given rise to a practice called "business continuity planning," which incorporates elements from disaster recovery planning and crisis management, including coordination of response to disruptions and maintenance of backup capacity for operational systems.[7]

Figure 2.3 shows the cyclical steps in the enterprise risk management process as well as the limitations of this process. The major steps are as follows:

- **Objective setting:** Management establishes the organization's risk appetite and risk tolerance and sets expectations for growth, profitability, and shareholder returns.

- **Event identification:** The potential threats or vulnerabilities as well as upside opportunities are identified for each line of business, and interdependencies are considered.

- **Risk assessment:** The likelihood and impact magnitude of each identified risk is estimated based on historical data and subjective assumptions, and management determines whether these risks are acceptable.

Figure 2.3. Limitations of enterprise risk management

- **Risk response:** Management considers the portfolio of risks in the context of its objectives. For risks that are deemed unacceptably high, measures are taken to reduce or mitigate the risks. Conversely, when the company posture is overly cautious, management may respond by taking on additional risk.

- **Control activities and monitoring:** Management implements controls to ensure that the identified risks are appropriately managed and monitored over time.

These ERM practices can help reduce both the likelihood and the severity of major incidents that can damage a company's reputation or profitability. Of course, these practices require constant vigilance and repeated updating to keep pace with changing conditions. Human error or omission is a frequent problem. After BP's Deepwater Horizon rig failed in 2010, spilling oil into the Gulf of Mexico, federal investigators stated that they "found no evidence that BP performed a formal risk assessment of critical operational decisions made in the days leading up to the blowout."[8] In contrast, companies that emphasize situational awareness and foresight have been able to anticipate and overcome major challenges, as illustrated in later chapters.

Limitations of Risk Management

The sequential process of ERM appears quite logical and thorough, but it is rooted in a simplistic, "reductionist" worldview. Each risk is identified and addressed independently, and hidden interactions are seldom recognized. The focus is on discrete events rather than gradual buildup of stresses. This procedural approach can lull the organization into a false sense of complacency that is shattered when an unexpected event occurs, as was arguably the case with the oil spill in the Gulf of Mexico. The complex, dynamic nature of global supply chains requires constant vigilance to sense potential vulnerabilities as well as exceptional agility and flexibility to respond to unexpected shocks.

Figure 2.3 shows several key limitations to the classic risk management paradigm:

- **Risks cannot always be anticipated.** A critical step in any risk management process is risk or hazard identification. Many of the risks that a company faces, however, are unpredictable or unknowable before the fact. Risks may not correspond to discrete events, but may result from cumulative changes that reach a tipping point. In a complex system, "emergent" risks are often triggered by improbable events whose causes are not understood, and their potential consequences are difficult to predict a priori. It would be impractical for companies to identify and investigate all the potential risks and vulnerabilities that may be hidden in their global supply chains.

- **Risks may be hard to quantify.** Even if risks can be identified, the lack of an adequate data set can make it difficult to assess the most significant threats. To assess the probability and magnitude of an identified risk, managers need reliable statistical information. Risk assessments are limited by the quality and credibility of the assumptions upon which they are based, and faulty assumptions or data may lead to misallocation of resources. That is especially a challenge in the case of low-probability, high-consequence events for which there is little empirical knowledge; in fact, managers may underestimate the probabilities or magnitudes of risks that they have never experienced.[9] One of the most difficult elements in risk assessment is the human factor. Human error or deliberate human malfeasance is a frequent cause of disruptions, but these triggering events are not easily modeled. Furthermore, as discussed in below, the presence of nonlinearities, cascading consequences, and interdependence among multiple threats requires a systemic rather than a reductionist approach.

• **Adaptation may be needed to remain competitive.** Risk mitigation and recovery practices, such as business continuity management, are typically aimed at returning to "normal" conditions. Instead, companies should strive to learn from disruptions and adapt their assets and business models to overcome potential weaknesses. Every disruption represents a learning opportunity and should be viewed as a stimulus to drive process improvement based on root-cause analysis and systems thinking. In today's fast-changing world, a philosophy of "business as usual" may be untenable. Companies that are quick to adapt may identify upside opportunities and seize competitive advantage.

The limitations of risk management have also been recognized in the regulatory environment. According to the National Academy of Sciences, risk-based methods are not adequate to address complex problems such as climate change and loss of biodiversity, and more sophisticated tools are available that go beyond risk management.[10] The concept of a stable equilibrium, with steady growth punctuated by occasional isolated deviations, is no longer realistic.

The established approaches of risk management can be very useful for protecting companies against predictable risks that are familiar and quantifiable, such as fires or power failures. However, they are not adequate for dealing effectively with the turbulence and complexity that characterize today's global risk landscape. The most damaging disruptions—as well as unexpected opportunities—tend to result from extreme events that are difficult or impossible to anticipate, let alone quantify.

Beyond Risk Management

Since 2000, the world has experienced a continuing stream of catastrophic events; examples range from the shocking September 11, 2001, terrorist attacks to the devastating earthquake in Haiti that took 160,000 lives in 2010. Catastrophes are often "black swan" events that have never been observed and seem implausible until they actually occur. It seems that we are often taken by surprise, although in hindsight we realize that we could have been better prepared.

Another wild card is technological innovation, which introduces disruptive changes that can completely overturn the risk and cost structure of entire industries. For example, the revolution in information and communication technology has resulted in a proliferation of smartphones that have more computing power than the mainframes of yesteryear. Such technology creates enormous upside opportunities for electronic commerce, but instantaneous mass communication has also produced destabilizing political forces and a variety of threats to company operations and reputation.

Every year in Geneva, the World Economic Forum assembles a group of experts to develop its annual *Global Risks* report.[11] This report identifies and analyzes a broad range of risk factors that may affect global economic development, from climate change to technological failures to political unrest. In recent years, the authors of this report have shifted from quantifying the relative likelihoods and consequences of specific risk factors to portraying the network of interdependencies among these factors (see figure 2.4).

The 2014 report acknowledges the importance of resilience for addressing "systemic" risks that are difficult to predict or manage effectively. Systemic risk is defined as "the risk of breakdowns in an entire system, as opposed to

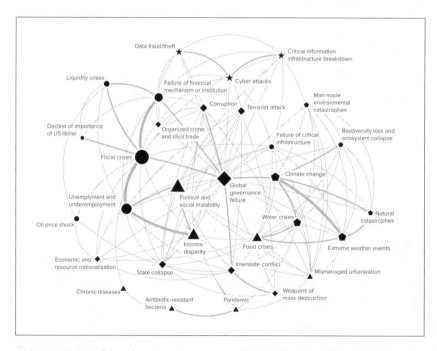

Figure 2.4. *Interdependencies among major global risk factors (from World Economic Forum)*

breakdowns in individual parts and components."[12] Systemic breakdowns can result from tipping points that trigger a chain of cascading effects, such as floods, power blackouts, property destruction, and economic crises.

The evolution of thought in the *Global Risks* report series reflects the increasing humility of managers confronted by a fluctuating risk landscape. In the face of complexity and turbulence, when disruptions are often unknowable and unforeseen, risk assessment becomes intractable, and traditional risk management practices are no longer adequate.

Risk management nevertheless remains an important methodology for dealing with recognized phenomena such as fires, accidents, diseases, and currency fluctuations. To address less tractable uncertainties, risk management needs to be supplemented with resilience management, which involves a different set of tools and metrics, based on systems thinking. In a constantly evolving global business environment, the notion of "optimization" is unrealistic; instead, companies need to adjust their risk posture dynamically in response to changing conditions.

In short, although ERM is a valuable practice that should not be abandoned, organizations need new strategies and more innovative approaches to deal with supply chain complexity and unexpected disruptions. Resilience thinking represents a fresh approach that can help overcome most of the above limitations and enable companies to cope more effectively with the daunting challenges of the modern risk landscape. Even insurance companies are recognizing the value of resilience; for example, Zurich Insurance adjusts its commercial insurance maximum loss estimates by a resilience factor that accounts for business continuity planning and ease of recovery.

As defined in chapter 1, enterprise resilience is the capacity to survive, adapt, and flourish in the face of turbulent change and uncertainty. A resilient company understands that managing uncertainty can lead to superior performance. Risk management tends to dwell on downside risks, but resilience thinking is equally relevant to upside rewards. Resilient companies are innovative and nimble in recognizing and capturing new business opportunities. As we shall see, the concept of resilience applies not only to enterprises, but to any self-organizing system. By learning from natural systems that have evolved for millions of years, enterprises can develop resilience strategies that provide competitive advantage.

Antifragile?

We live in a culture that values order and predictability. As a result, most of our artifacts and institutions are fragile and are easily damaged by random forces. What if an object were antifragile; that is, what if it actually thrived on chaos?

That is the fascinating premise of *Antifragile: Things That Gain from Disorder*, a book by Nassim Nicholas Taleb.[13] In his previous book, *The Black Swan*, he pointed out the futility of trying to predict major disruptive events (e.g., recessions, revolutions, disasters) with cascading consequences that could change the course of our lives. In this sequel, he argues that we should accept uncertainty as not only inevitable, but even beneficial; after all, biological organisms can adapt and regenerate in response to random shocks or fluctuations. Stress is an essential aspect of life, and it makes you stronger.

Taleb, a former businessman turned philosopher, proposes a fundamental fragile-robust-antifragile triad, a sort of spectrum along which everything can be positioned. The systems that we design to be robust are actually vulnerable to unexpected events or forces. Antifragility goes beyond robustness in that it benefits from disorder.

Taleb is merciless in skewering the "fragilista," those who cling to the illusion of order and predictability, including government bureaucrats, bankers, physicians, and even fitness trainers. He calls risk a "sissy" concept and is openly scornful of academics who pursue reductionism and elimination of uncertainty. Instead, he advocates "decision making under opacity."[14] Although his views are extreme, he presents a provocative challenge to the conventional wisdom of risk management.

Unfortunately, *Antifragile* conflates resilience and robustness, treating them as synonymous. In practice, the meaning of resilience is actually very close to the notion of antifragility. Rather than resisting change, resilient systems are able to survive, adapt, and flourish in a volatile environment.

Resilience in Action

Climate Adaptation by Entergy and Swiss Re[15]

Although debates over climate change may linger, some companies are taking positive action to understand potential climate risks and position themselves accordingly. One example is Entergy, an electric utility company operating in the United States, including along the Gulf Coast. In 2010, Entergy partnered with America's Energy Coast and America's Wetlands Foundation to quantify climate risks in this region and identify economically sensible approaches for building a resilient Gulf Coast.[16] The global reinsurance company, Swiss Re, was a lead contributor to this study and applied a methodology called "economics of climate adaptation" to build a portfolio of economically suitable adaptation measures.

This study represents the first comprehensive analysis of climate risks and adaptation economics along the US Gulf Coast. The study team's projections were sobering. They estimated that over a twenty-year time frame, from 2010 to 2030, annual economic losses due to extreme storms would increase by 50 to 65 percent, resulting in more than $350 billion of cumulative expected losses. This figure includes about 7 percent of total capital investment for the Gulf Coast area and 3 percent of annual gross domestic product (GDP) that would go toward reconstruction activities. Severe hurricanes such as Katrina could also have a significant dampening effect on growth and reinvestment in the region.

Ideally, the Gulf Coast needs to identify a portfolio of adaptation solutions that involve "no regrets"; in other words, the solutions should have low investment needs, high potential for loss reduction, and additional significant benefits (e.g., wetlands restoration). Such investments will avoid mortgaging the future in the sense of imposing a heavy financial burden with an uncertain payback.

The study methodology involved the following steps.

1. **Hazard assessment:** Three key hazards were considered: hurricanes, subsidence of land, and sea-level rise. Future scenarios were developed in consultation with expert scientists in the

field. There is broad agreement that warmer sea-surface temperatures in the future may lead to more severe hurricanes. To address uncertainty in climate change, three scenarios were developed in the 2030 and 2050 time frames, representing low, average, and extreme climate change. Natural hazard modeling, using probabilistic simulation of tropical cyclones and hurricanes, was done in collaboration with Swiss Re.

2. **Economic value assessment:** This step required estimating the size and location of current and future assets along the Gulf Coast, accounting for both the replacement value of physical assets and the economic value of business interruption. The Gulf Coast currently has more than $2 trillion dollars in asset value and is expected to grow to more than $3 trillion in the 2030 time frame. The analysis included a detailed and granular assessment of oil and gas assets and electric utility assets, covering more than 50,000 pipelines, offshore structures, and wells; more than 500,000 miles of electric transmission and distribution assets; and about 300 generation facilities.[17]

3. **Vulnerability assessment:** Vulnerability curves were developed relating the value at risk to events of different severities. A vulnerability curve shows the correlation between hurricane severity (i.e., height of storm surge, wind speed) and asset loss (i.e., percentage lost of total asset value). Different categories of assets typically have different vulnerability curves; for example, residential property may be quite different from utility assets in terms of vulnerability to extreme winds. Similarly, within a single asset category, vulnerability curves are highly sensitive to parameters such as construction codes or materials used.

The analysis concluded that the Gulf Coast faces significant losses today, averaging on the order of $14 billion per year. These losses are expected to increase going forward, amounting to about $18 billion per year (with no climate change) or $23 billion per year (with extreme climate change) by 2030. Losses may increase further by 2050, ranging from about $26 billion to $40 billion per year. Current loss rates represent about 2 to 3 percent of the region's

GDP and about 7 percent of the region's annual capital investment. The implication is that the Gulf Coast region spends about 7 percent of its invested capital each year on rebuilding infrastructure rather than on capital investments that could be driving future economic growth.

A key finding of the study is that *regardless of climate change*, the Gulf Coast faces an increase in risks from natural hazards going forward. Approximately half the increase is driven entirely by baseline economic growth and subsidence unrelated to climate change. Among economic sectors, the oil and gas industry and commercial/residential interests were found to be particularly vulnerable, accounting for about 88 percent of loss in the 2030 time frame.

To prevent or mitigate these anticipated losses, a broad range of potential measures were identified, ranging from infrastructure upgrades to systemic behavior change to risk transfer via insurance. The primary focus was on measures that can be implemented today, so that future innovations in technology (e.g., hurricane-resilient building materials and methods) were not assessed. Therefore, the analyses should be repeated periodically to account for technological innovations.

The study then considered both the costs and benefits of the adaptation measures to help prioritize actions. The cost analysis considered the present value of life cycle costs over time, including capital expenditures, operating expenditures, and operating expenditure savings. Similarly, the benefits analysis considered the present value of averted losses over time. It was found that, in the near-term, potentially attractive measures can address almost all the increase in loss and thus maintain a constant risk profile for the region.

The study team concluded that investing in measures totaling about $50 billion over the 2010–2030 time frame would lead to approximately $135 billion in averted losses over the lifetime of those measures. On a broader scale, pursuing all potentially attractive actions would involve an investment of approximately $120 billion over that time frame and may lead to $200 billion in averted losses. The portfolio of measures would include a focus on

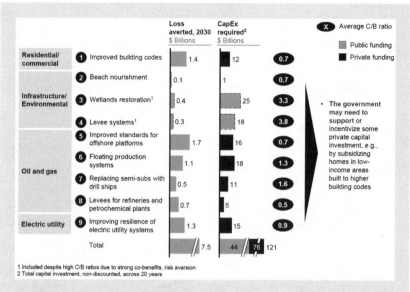

Figure 2.5. *Recommended climate adaptation measures and estimated benefits*

adaptation to address near-term risks combined with mitigation to address longer-term risks. The study did not, however, try to monetize the additional co-benefits that will result from these measures; for example, wetlands protection and restoration will help ensure the resilience of critical ecosystem services (see chapter 7).

Figure 2.5 depicts a grouping of the recommended measures into nine broad categories across all sectors: residential/commercial, infrastructure/environmental, oil and gas, and electric utility. The study recommended that approximately $44 billion of public funding be invested from 2010 to 2030 to fund key infrastructure projects (including wetlands and levees). In addition, some $76 billion in private funding would be required. Policy makers, however, may need to support and provide incentives for some private capital investment, such as by subsidizing homes in low-income areas built to stricter building codes.

The most cost-effective way to offset the remaining $14 billion in annual expected losses associated with extreme events is through insurance or risk transfer. Four key risk transfer actions can help address residual loss: increasing penetration of existing insurance through more affordable premiums linked to physical measures, decreasing the prevalence of underinsurance through incentives that encourage updating of insured value of property, encouraging additional self-insurance, and transferring top-layer risk (e.g., through catastrophe bonds).

The recommended climate adaptation measures will require cooperation among a broad set of stakeholders with conflicting interests and varying levels of effectiveness. For example, measures related to improved construction codes may require new policies to be put in place by local regulators, investments to be made by individual home owners, and appropriate enforcement. The aim of the study was to develop practical solutions that will take Gulf Coast resilience to the next level. Although significant and broad stakeholder engagement will be required, these actions are essential to place the region on a resilient path going forward. The alternative will be to enter a long-term spiral of increasing losses with corresponding adverse economic and social impacts.

In a 2014 interview, David Bresch, head of sustainability and political risk management at Swiss Re, commented on the Entergy climate adaptation study:

> At Swiss Re we have realized that the real opportunity for Entergy and other firms goes beyond avoiding damage; it is about being better organized to rapidly detect problems and to best serve their customers. This way, they'll gain market share after an event, as it will take competitors longer to be back in business. It is not necessarily about protecting an electric substation, it is more about how you engage with customers to be prepared for an emergency. Serving a customer by bringing in a generator may succeed just as well as flood-proofing a substation, and may be a more economical use of resources. The question is: what helps to best serve the joint interests of both the company and its customers? This study provided good insights into how companies can organize to weather these types of challenges and improve their resilience.

Takeaway Points

- In the world of operations, risk is an undesirable threat to be avoided; in the world of finance, risk is a fundamental uncertainty that can be managed; that is, without risk, there is no reward.

- These different views of risk are reconciled under enterprise risk management, which enables corporations to gauge the appropriate level of risk in an uncertain business environment.

- Conventional risk management is helpful for familiar threats but has severe limitations in a world of turbulent change and unforeseen black swan events.

- Resilience thinking augments risk management by helping companies cope with turbulence, respond effectively to unforeseen disruptions, and adapt to change.

- **Resilience in Action:** Entergy worked with Swiss Re and others to analyze the potential long-term losses associated with climate change in the Gulf Coast region and recommended several cost-effective adaptation strategies to improve overall resilience.

CHAPTER THREE

Systems Thinking

We can't impose our will upon a system. We can listen to what the system tells us, and discover how its properties and our values can work together to bring forth something much better than could ever be produced by our will alone.

Donnella Meadows[1]

Every company is a complex system, interacting with other systems in a constantly changing environment. The more we understand this fact, the more humble we become about our ability to control these systems. Embracing change will involve a true paradigm shift, from preserving system stability and minimizing deviations from "normal" to accepting the inevitability of change and maximizing adaptive capacity. Enterprises that are accustomed to traditional management styles will need to reexamine and possibly modify their cultural norms and business processes. The question of how to build organizational resilience is explored further in chapter 8.

The Illusion of Control

Because humans seem to crave stability, we can easily be lulled into a belief that the world is orderly and predictable, but preserving order is a constant effort and requires significant investments of energy and other resources. If we turn off the power or discontinue the maintenance, our marvelous machines are soon rendered useless. The universe outside our gates, including the natural environment, is infinitely complex, dynamic, and variable. The notion that we are in control can be a dangerous illusion, and every so often a natural or man-made disaster reminds us of our vulnerability.

In fact, most businesses today operate in a highly networked economy, with supply chains and trading relationships that extend around the world. This connectivity on a global scale creates many opportunities for economic efficiency and mutual advantage, but it also exposes companies to a variety of global disruptions. A minor glitch in a microchip plant in Taiwan can alter production schedules and costs for global electronics manufacturers and can negatively affect consumers. This network of interdependencies makes an enterprise both more robust and more fragile; it is robust in the sense of collective capacity, but fragile in the sense of collective vulnerability.[2]

The conventional risk management practices described in chapter 2 are well suited to a relatively steady business environment with known risks. In such a case, it is possible to measure risk likelihoods and magnitudes based on past experience and to make decisions about managing and mitigating the high-priority risks. When risks are unknown, however, companies must rely on learning and adaptation to cope with unexpected disruptions. Moreover, as volatility increases, companies need to be prepared for risks or opportunities that may emerge in a seemingly random fashion. A wise strategy for managing a complex system in today's turbulent environment is to develop *inherent* resilience: the capacity to survive, adapt, and flourish in the face of shocks and stresses that may be unfamiliar and unexpected.

Disruptions are not always triggered by catastrophic events. In complex supply networks, small disturbances can occasionally cascade into massive discontinuities that have lasting effects on the business. As described in chapter 6, a 2002 labor dispute in California shut down West Coast ports for several weeks, costing US companies roughly $1 billion per day. Nonlinearity implies that these radical shifts can occur suddenly, when conditions reach a tipping point, as when long-simmering tensions finally ignite a revolt. Unfortunately, the very system complexity that generates these disturbances makes it virtually impossible to predict their nature or timing. Smooth changes can usually be tolerated by adjusting the system behavior, but *real systems don't have smooth curves*. Although it is difficult to predict rare events, we may be able to improve enterprise resilience by anticipating the forces of change and finding creative ways to take advantage of the system dynamics rather than merely reacting to disturbances.

Taking a Systems Approach

Managing enterprise resilience requires a radical change in mind-set, which may be difficult for those accustomed to a tightly focused, result-oriented approach. Instead of concentrating on specific "things"—products, vehicles, people—the practice of *systems thinking* strives to develop a broader understanding of the dynamic relationships among those things. Taking a systems approach can help us question conventional wisdom and develop profound new insights. Of course, to be used effectively in a management context, systems thinking must be linked with concrete decision making.

Perhaps the best-known advocate of systems thinking is Peter Senge, who popularized the concept of a learning organization where "people continually expand their capacity to create the results they truly desire, where new and expansive patterns of thinking are nurtured, where collective aspiration is set free, and where people are continually learning to see the whole together."[3] Learning organizations are able to achieve excellent performance by being flexible, adaptive, and productive in situations of rapid change. People can learn to manage highly complex systems by combining an intuitive holistic understanding with a rational and purposeful approach.

A *system* can be defined as an interrelated set of components that form a structure and perform a function. For example, the core of an enterprise system consists of business divisions, functional departments, and company-owned assets. An extended view of an enterprise system might include suppliers, customers, and other stakeholders as well as the public infrastructures that support business operations (see chapter 11).

More broadly, examples of systems range from biological systems (e.g., ant colonies) to engineered systems (e.g., electrical power grid) to social systems (e.g., professional networks). Systems theory is the study of how these complex entities interact with their environments and evolve by acquiring new, emergent properties.[4] Rather than characterize a particular entity (e.g., a company) by the properties of its parts or elements (e.g., plant capacities), systems theory focuses on the relationships between the parts that connect them into a whole. This perspective can reveal whole-system properties (such as resilience) that cannot be understood merely by analyzing the parts.

Complex systems are generally dynamic, nonlinear, and capable of self-organization to sustain their existence. They are able to achieve dynamic equilibrium through feedback loops; for example, by pollinating flowers, bees create a feedback loop that reinforces the production of nectar. Similarly, by supporting social and philanthropic activities, corporations strengthen the vitality of the communities to which their employees belong. Companies like Wal-Mart and Microsoft have been compared to an ecological *keystone* species, occupying the hub of a business network and improving the overall health and robustness of the network.[5]

An understanding of system resilience becomes important when significant disruptions, or discontinuities, shift the system away from its current equilibrium state. Such disruptions could include the introduction of new technologies, the emergence of new regulatory and market forces, or changes in the availability of resources. For example, will the broad introduction of low-cost electronic devices in developing nations create an excessive flow of postconsumer wastes, and if so, how will local governments respond? Some believe that emerging sustainability issues will catalyze innovations that will change the basis of competition in many industries.[6] Those that wish to reap the benefits of these changes must also be alert to the risks, since complex systems may be vulnerable to small, unforeseen perturbations that cause catastrophic failures. Risk management practitioners are adopting resilience concepts as a way to "withstand systemic discontinuities and adapt to new risk environments."[7]

It is important to distinguish systems thinking from the established practice of *systems engineering*, a discipline developed during the 1940s to design and manage complex, engineered systems such as bridges and military aircraft. Systems engineering provides a rigorous, hierarchical approach for coordinating a complex project, including the design and acquisition of components and subsystems as well as quality assurance, reliability testing, logistics, and resource needs over the full asset life cycle. Although systems engineering does take an integrated view of the system, it has been dominated by a traditional engineering mind-set, emphasizing precision and control over resilience and adaptability. Every component of the system must operate as specified for the system to work correctly. This view stands in contrast to fault-tolerant design principles that allow for systems to continue functioning or to degrade gracefully in the event of failures. Systems engineering tends to focus on optimizing cost and performance, and it views external systems as constraints or boundary conditions rather than considering their dynamic behavior.

Enterprise Connectivity and Hierarchy

To adopt a systems view of the enterprise, we need to consider the dual characteristics of *connectivity* and *hierarchy*. In the ecological realm, all living creatures are connected and are part of a hierarchy of nested systems, or layers. An individual creature is linked to plants, predators, prey, symbiotic creatures, and its overall habitat. It is also part of a layered hierarchy that extends from a family to a social group up to a larger community, and it is composed of layers that extend from organs to tissues down to individual cells. These same characteristics are observed in the business world.

Business *connectivity* arises from strategic partnerships, joint ventures, and extended supply chains that couple suppliers and customers within a "virtual" enterprise. More broadly, businesses are connected to external stakeholder groups that can influence shareholder value. Examples of influential factors include employee attitudes, consumer preferences, regulatory policies, community interests, and competitor initiatives. To the extent that political or economic changes can destabilize these stakeholder systems, enterprise performance can be severely affected. For example, shareholder resolutions, labor strikes, regulatory actions, community protests, or the introduction of new technologies can disrupt or derail a business.

Business *hierarchy* arises from the structural layers that typify the modern enterprise. They include organizational hierarchies (department, business unit, enterprise), product hierarchies (model, family, brand), and process hierarchies (unit process, facility, supply chain). Figure 3.1 presents a simplified view of three layers, depicting the business unit, enterprise, and global levels. Each layer consists of dynamic systems that fluctuate over time, and each involves linkages among economic, social, and physical systems. Disruptions in any system within a given layer of the

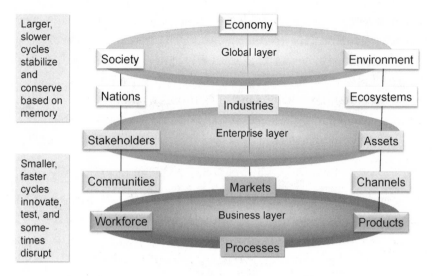

Figure 3.1. Enterprise system linkages and layers

hierarchy can have cascading effects on the resilience of other systems in multiple layers.

Typically, these interconnected systems experience cycles of change that operate at different speeds. Global changes can occur over periods of years or centuries, whereas business-level changes can occur in days or hours. At the higher levels, slower cycles have a conservative, stabilizing effect based on memory of past cycles; for example, constitutional principles at the national level provide a mechanism for resisting rapid change. The lower levels tend to have smaller, faster cycles that innovate, test, and sometimes disrupt the existing equilibrium. For example, the emergence of the Internet in the 1990s transformed both social interactions and commercial transactions in the space of a decade.

This hierarchical view leads to the following "nested" interpretation of resilience.

- A resilient *society* continues to grow and evolve to satisfy the changing needs and expectations of its population, including present and future generations.

- A resilient *enterprise* continues to grow and evolve to meet the changing needs and expectations of shareholders and stakeholders in its host society.

- A resilient *business unit* continues to grow and evolve to meet the changing needs and expectations of its parent enterprise, its markets, its workforce, and its business partners.

- A resilient *product, process,* or *supply chain* continues to evolve to meet the changing needs and expectations—including cost, performance, and service—of its parent business, supply chain participants, customers, and other stakeholders.

A process cannot be resilient in an absolute sense; rather, it must be considered in the context of the supply chain, the market, and the broader environment. Therefore, designing a resilient enterprise requires an understanding of how products, processes, and services interact with these external systems. In a sense, product and service systems are the offspring of corporations and acquire important characteristics from their "parents," including branding, technology, distribution channels, and stakeholder perception. Today, corporations are increasingly expected to disclose the details of their supply chain processes, including raw materials and the conditions under which they were manufactured. Do they employ forced labor or child labor? Do they use recycled material content or renewable energy? Were chlorofluorocarbons used in the manufacturing process? Do their products contain genetically engineered constituents? Vulnerabilities in any of these aspects can weaken the enterprise as a whole; conversely, the resilience of the enterprise depends on the collective capabilities of its many components.

The Triple Value Framework

One approach that has proved useful for systems thinking is the triple value (3V) framework, which represents the linkages and flows of value among industrial, societal, and environmental systems.[8] It has been used for characterizing regional sustainability and resilience issues and for exploring the possible consequences of alternative "interventions," including industrial policies and management practices. Figure 3.2 presents a simplified view of the 3V framework at a national or global scale, partitioning the world into three types of interconnected systems. The 3V framework is based on a conceptual framework originally developed for the Organization for Economic Cooperation and Development under its Sustainable Materials Management initiative.[9] There, the framework

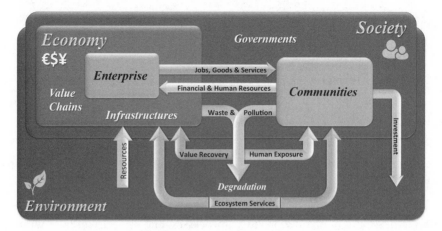

Figure 3.2. *The triple value framework, global view*

was used to analyze the consequences of innovation in global material life cycles, including material extraction, processing, transportation, use, and disposal.

Today, it has become evident that global changes and interactions among technology, geopolitics, and the environment can disrupt cycles of material and energy flows. For example, few people foresaw that corn-based ethanol production in the United States would drive up food prices in Mexico, that floods in the Mississippi basin could cause fuel shortages, or that the need to supply food to a growing world population would exacerbate imbalances in the global nitrogen cycle. Traditional approaches to economic and environmental management, however, are based on static, compartmentalized models that naïvely assume a steady-state equilibrium. In truth, the natural and industrial systems that we try to "manage" are tightly coupled, dynamic systems operating far from equilibrium, exhibiting nonlinear and sometimes chaotic behavior.

A systems approach is needed to characterize the interdependencies and feedback loops described above. Such an approach should include the flows of information, materials, energy, financial capital, and labor among economic systems (e.g., extraction, agriculture, and manufacturing), societal systems (e.g., urban centers, education, communication, and governance), and natural systems (e.g., air, water, soil, and ecological systems). The 3V framework captures the relationships among these three types of systems as follows.

- **Economic systems** use both human and environmental resources to fulfill societal demands. Companies extract or "harvest" resources from the environment, including energy, materials, water, and food; add economic value through supply chain operations; deliver products and services to societal markets; and deposit industrial wastes into the environment. Their productive capacity is embodied in both **built capital**, including infrastructure, and **intellectual capital**, including industrial technologies. The flow of commercial products and services provides value to consumers, while corporate profitability contributes to shareholder value, creates jobs, and improves the economic prosperity of communities.

- **Societal systems**—that is, human communities and institutions—consume the products, services, and energy supplied by economic systems and generate wastes that may either be recycled for value recovery or deposited into the environment. Societies may also benefit directly from natural amenities, such as clean air, clean water, recreational uses, and psychic enjoyment. Human health and well-being may be affected beneficially or adversely by changes in economic or environmental systems, such as avoidance of hazardous waste generation. **Human capital and social capital** deliver economic value to industry by providing essential workforce skills and market stability. In addition, societal systems provide public institutions and governance mechanisms that guide human behavior and, in particular, generate environmental value by protecting and restoring environmental resources.

- **Environmental systems** are the ultimate source of all material and energy resources. They contain reservoirs of natural resources, including renewable resources (e.g., forests) that can be replenished over time; nonrenewable resources (e.g., petroleum); and finite environmental media (e.g., air, water, and land) that may become degraded. The productive capacity of environmental systems is known as **natural capital**,[10] and the flow of ecosystem goods and services delivers value to both industrial and societal systems.

Economic and societal systems are coupled by economic market transactions that mediate the flows of goods and services. Economic growth generally corresponds to increased material throughput and usually correlates with population growth, increased wealth, and growth in demand per capita. The linkage between economy and society is a **positive feedback loop**, and without external constraints, it theoretically enables perpetual growth with increasing human prosperity.

On the other hand, the linkages of these systems with the environment represent a **negative feedback loop**. We are consuming ecosystem services faster than we can replenish our natural capital, which includes freshwater, soil, forests, coral reefs, and glaciers. To make matters worse, we are generating large amounts of waste and emissions that degrade these very ecosystems. Increasing economic and population growth, coupled with increasing waste and emissions, will eventually overwhelm the capacity of the planet to service human needs. When markets fail to account for economic externalities such as gradual degradation of soil and water quality, the result is a loss of opportunity for future generations, sometimes called an "intertemporal market failure."[11] Many resource economists argue that we can prevent such market failures by replacing traditional GDP with more comprehensive indicators such as "inclusive wealth" measures, which explicitly assign value to social and natural capital. One possible solution is "dematerialization": reducing the volume of material throughput required to achieve economic growth and prosperity. For example, as described in chapter 4, economic growth can be decoupled from material throughput by creating a closed-loop "circular economy" in which waste materials are recovered and converted into feedstocks rather than being sent to landfills or incinerated.

The 3V framework resembles the "triple bottom line" concept that is often invoked in sustainability discussions, but it goes much further by explicitly showing the interdependencies and value drivers among the three systems: industry, society, and environment. The bottom-line metaphor is incomplete; chapter 5 describes the different pathways whereby companies generate shareholder value: tangible financial returns, enhancement of intangible assets such as reputation and human capital, and delivering value to stakeholders, which indirectly strengthens those intangible assets.[12] The financial bottom line is only a measure of cash flow and does not reflect the importance of capital preservation and renewal for value creation. Moreover, accounting separately for economic, social, and environmental performance fails to recognize the inherent synergies among these three dimensions. For corporations, there is truly only one bottom line, and the benefits derived from economic, social, and environmental performance are blended into value creation for both company shareholders and society at large.

Because the 3V framework represents the dynamics of resource flows among economic, social, and environmental systems, it is useful for understanding enterprise resilience. Ideally, these systems can achieve a dynamic equilibrium in which the flows of material and energy resources are sustainably balanced with the health and vitality of social and natural capital. Such equilibrium can be achieved by introducing positive feedback loops to offset the depletion of natural resources. For example, the concept of a circular economy suggests that wastes have residual value and can be recovered and reused instead of being released into the environment. On a broader scale, companies and communities can invest in environmental protection and restoration, thereby protecting or enhancing the availability of natural capital.

In reality sudden disruptions or gradual shifts can delay or derail progress toward sustainability and can destabilize the balanced operation of these systems, leading to economic stagnation, social deprivation or environmental degradation. Therefore, it is important to build resilience into these systems to alleviate stresses and limit the potential for damaging shocks. Every pathway of value flow in figure 3.2 represents a potential vulnerability. For example:

- Natural resources such as water, fuels, minerals, and biomass are needed as feedstocks for energy generation, manufacturing, agriculture, and other economic activities. Resource scarcity or infrastructure breakdowns can threaten the availability of energy, water, transportation, and other critical services.

- Human resources and innovation are needed to support continued economic development and industrial competitiveness. There are a variety of disruptive social forces, including political upheaval, social unrest, poverty, corruption, and lawlessness, that can threaten the stability of economic activities and jeopardize quality of life.

- Population growth and the transition of developing nations to a more affluent lifestyle create pressures on natural capital. At the same time, inefficient use of resources leads to excessive discharges of wastes and emissions, contributing to climate change, sea-level rise, ecosystem degradation, and threats to human health and safety.

- Urban communities are particularly vulnerable to sudden disruptions caused by a confluence of the above forces. For example, there is great concern about potential disruptions that can have cascading effects due to the food-energy-water nexus: the interdependence among municipal infrastructures and the value chains that support them (see chapter 4).

There are no quick and dirty solutions to these pervasive concerns. To maintain their competitiveness, enterprises need to tackle the challenge of understanding the dynamics and potential vulnerabilities of these coupled systems, assessing the relative importance of possible disruptions from an enterprise perspective, developing effective strategies for improvement and adaptation, and designing for inherent resilience.

Resilience in Action

Applications of Systems Thinking

Systems thinking provides a holistic approach for understanding the dynamic interactions among complex economic, environmental, and social systems and for evaluating the potential consequences of interventions, such as new policies, new technologies, and new operating practices. Application of systems thinking often involves the use of *system dynamics* models to characterize the interdependence, feedback loops, and dynamic behaviors of complex, adaptive systems.[13] Two examples of system dynamics modeling applications based on the 3V framework are discussed below. Figure 3.2 depicts the very highest level of system aggregation, corresponding to the topmost global layer of figure 3.1. The 3V framework enables drilling down to more granular levels with a narrower focus on specific companies, industries, geographic regions, and communities. It also enables vertical linkages; for example, global economic trends may influence local markets.

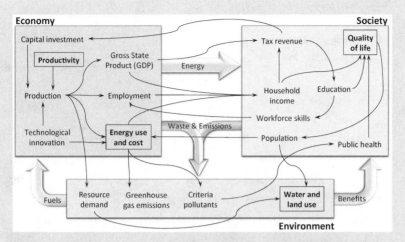

Figure 3.3. *Overview of key variables and system linkages in the dynamic energy–economic policy simulation (DEEPS) model*

One example of how 3V was applied for strategic policy development is shown in figure 3.3. In 2009, the State of Ohio commissioned a study of economic and energy resilience in the face of potential federal actions to limit global warming emissions. Ohio relies heavily on fossil-fuel-based electric power, and many of the state's industries, such as steel and glass manufacturing, are highly energy-intensive. The purpose of the study was to develop a dynamic energy–economic policy simulation (DEEPS) to assess state-level energy policy options for assuring the continued competitiveness of Ohio industries under various future scenarios.[14]

Unlike traditional econometric models, which use statistical tools to generate future projections based on historical patterns, the DEEPS logic is based on cause-and-effect relationships between key variables. For example, population growth tends to increase energy demand, whereas technological innovation tends to reduce the cost of energy generation as well as the associated environmental impacts. Economic growth increases tax revenues and household income, which tends to improve education and workforce skills. The feedback loops among these and other causal chains will drive future energy costs for consumers as well as secondary consequences for quality of life and environmental protection. The model indicates that investment in energy efficiency and renewable energy technologies, despite their higher initial cost, will stimulate the Ohio economy, reduce overall energy expenditures, and provide greater resilience against service interruptions due to the diversity of energy sources.[15]

Another example of systems thinking at a watershed scale is a collaboration between the US EPA and several government and business organizations in southern New England, encompassing Rhode Island and parts of Massachusetts. As in many regions of the United States, excessive releases of *nutrients*—mainly nitrogen and phosphorus—from wastewater, agriculture, and stormwater run-off are causing algae blooms that degrade aquatic ecosystems and impair water quality, sometimes even resulting in isolated fishkills. Nutrient pollution sets up a tension between population growth, urban development, and agricultural production on the one hand and the interests of local citizens and key New England industries such as fishing, recreation, and tourism on the other hand. Meanwhile, these problems are being further aggravated by effects of climate change that are already evident, including rising sea levels, increasing storm intensity, and coastal flooding.

The collaborative team, supported by EPA's Office of Research and Development, has been exploring how a systems approach can help anticipate change and address the complex problem of sustaining economic growth while protecting precious water resources and adapting to climate change. In 2013, a triple value simulation (3VS) model was developed for the Narragansett Bay watershed—which includes the cities of Providence and Newport, Rhode Island—to evaluate alternative strategies for coastal sustainability and resilience.[16] The 3VS model, illustrated in figure 3.4, is designed to help policy makers and stakeholders develop robust solutions, taking into account projected increases in urban population and climate effects. By evaluating key indicators such as nutrient concentrations, beach visits, and tourism revenue, the model has shown that traditional point-source controls such as advanced wastewater treatment can

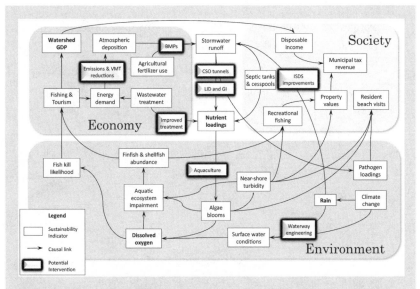

Figure 3.4. *Overview of system dynamics and potential interventions in the Narragansett Bay triple value simulation model*

be supplemented by alternative, affordable technologies and practices. One example of a beneficial intervention is the use of natural landscapes—known as *green infrastructure*—to attenuate nutrient flows, protect waterways from storm runoff, improve ecosystem health, and reduce the effects of coastal storms. Chapter 7 describes how Dow Chemical has deployed green infrastructure as an alternative to traditional wastewater management.

Takeaway Points

- Enterprises are complex systems that interact continuously with external systems in a dynamic business environment, and many of the factors that influence resilience are outside the control of the enterprise.

- Preserving order in a volatile world is an ongoing challenge; rather than merely reacting to disruptions, enterprises can often anticipate change based on insights about possible future conditions and system behaviors.

- Taking a systems approach helps enterprise managers understand interdependencies among economic, social, and environmental systems at many levels, from a specific business process to the global economy.

- The triple value (3V) framework provides a generic model of the dynamic linkages and interactions among economic, social, and environmental systems and reveals vulnerabilities that may cause disruptions.

- **Resilience in Action:** Applications of 3V include anticipating federal energy and environmental regulations to ensure the resilience of industrial systems in the state of Ohio, and developing strategies to protect coastal water resources for fishing and tourism in southeastern New England.

CHAPTER FOUR

The Resilient Enterprise

Resilience thinking is structured around the acceptance of
disturbance, even the generation of disturbance, to give a
system a wide operating space.

Brian Walker and David Salt[1]

The law of entropy tells us that everything in the universe is descending into chaos, unwinding like a clock. Living things, including human beings, are engaged in a constant struggle to maintain order and structure. The essence of life is resisting entropy, or creating order out of chaos. Living things do not just react to their environment; rather, they participate in complex feedback loops that shape their environment. Mammals, fish, and insect colonies build elaborate physical structures and social networks that enable them to replicate and flourish. Likewise, business enterprises and human communities create orderly structures and networks that enable them to grow and flourish. Here's the problem, though: The extraordinary success of humans may come at the expense of natural systems, ironically undermining our own resilience.

Lessons from Living Systems

Living things are inherently resilient because they are able to adapt to disturbances. Networks of living things are even more resilient than individual living things, although they are also vulnerable to catastrophic failures such as disease epidemics. Science teaches us that nature is resilient at every level, from the functioning of an individual cell to the evolution of a species to the intricate balance of a food web. Living systems sense emerging threats, protect their vital assets, heal their wounds, modify their structures, and adapt to environmental changes. Of course, living systems require a continuous supply of energy, mainly from the sun. There are many examples of how living systems have responded to unexpected challenges, recovered from disruptions, and continued to grow and regenerate.

- In the early 1990s, Iraqi dictator Saddam Hussein drained the extensive Tigris-Euphrates wetlands known as the Mesopotamian marshes, partly to punish the indigenous Arab tribes who opposed his regime. After the fall of Hussein in 2003, water was allowed to flow into the parched land, and within months, new vegetation and bird life began to flourish again.

- In 2001, in Edmonton, Alberta, a thirteen-month-old girl crawled out of her home at night in subzero weather. Her mother found her frozen body curled up in the snow, clad only in a diaper. The girl was rushed to a local hospital and declared clinically dead, but miraculously, as she warmed up, her heart restarted spontaneously, and she came back to life.

- In 2007, the small city of Greensburg, Kansas, was devastated by a tornado that killed eleven people and destroyed 95 percent of existing structures. Rather than abandoning the town, the citizens worked together to rebuild it as an ecologically conscious city, featuring green buildings and wind energy. Greensburg has now become famous as a model of a sustainable living community.

Enterprises are very much like living organisms and often exhibit similar behaviors. As mentioned in chapter 1, however, the engineered systems that enterprises create are generally brittle in contrast to living systems. Brittleness can be seen at the level of a simple mechanical device or a complex system such as an aircraft. Engineered systems are designed for specific operating conditions and must be regularly monitored and maintained to ensure that these conditions are met; otherwise, they are liable to fail. Thus, they are poorly equipped to deal with unexpected challenges and are unlikely to recover from nonroutine disruptions. Efforts to design "intelligent" robotic systems with greater adaptability are promising, but they remain far more primitive than even simple organisms.

Some engineers and designers have begun to practice *biomimicry*, taking inspiration from designs that have evolved in nature based on millions of years of evolution.[2] Examples include the design of swimsuits that mimic the tiny scales on sharkskin and the design of Velcro based on the spikes of burrs. At a more macro level, many manufacturing firms have introduced *industrial ecology* practices, developing closed-loop solutions for beneficial reuse of waste materials.[3] These practices could be called "eco-mimicry" because they are inspired by the cyclical patterns of material flow in nature, where there is no such thing as waste. Simple examples are common. Cement manufacturers incinerate contaminated wastes in cement kilns as a substitute for fossil fuel, effectively reducing global warming emissions; electric utilities recycle fly ash from their boilers for blending into cement and capture waste heat for local applications; and electronic equipment recyclers recover parts and materials from discarded devices and recycle them into a variety of secondary uses. Going further, the Biomimicry Institute has formulated a set of life's principles that encourage us to learn from nature's patterns (see "Life's Principles" box).

Life's Principles

Evolve to Survive

- Replicate strategies that work.
- Integrate the unexpected.
- Reshuffle information.

Be Resource (Material and Energy) Efficient

- Use multifunctional design.
- Use low-energy processes.
- Recycle all materials.
- Fit form to function.

Adapt to Changing Conditions

- Maintain integrity through self-renewal.
- Embody resilience through variation, redundancy, and decentralization.
- Incorporate diversity.

Integrate Development with Growth

- Combine modular and nested components.
- Build from the bottom up.
- Self-organize.

Be Locally Attuned and Responsive

- Use readily available materials and energy.
- Cultivate cooperative relationships.
- Leverage cyclic processes.
- Use feedback loops.

Use Life-Friendly Chemistry

- Build selectively with a small subset of elements.
- Break down products into benign constituents.
- Do chemistry in water.

In addition to mimicking nature, companies need to be aware of their impacts on nature. Scientists may never be able to predict accurately how industrial activities will affect biological or ecological systems. It is often said that nature is resilient. Indeed, natural systems have the capacity to tolerate perturbations, to recover slowly from severe damage, and to evolve into new, unimagined forms. Some scientists, however, warn that we have already entered a situation of "overshoot," meaning that our demands on natural resources have overshot the global capacity of the planet.[4] Therefore, it is important to understand the properties of both natural and engineered systems that make them more or less resilient. For example, many ecologists believe that a decrease in biodiversity will tend to reduce ecosystem stability.[5] Similarly, it has been found that a lack of managerial diversity reduces a company's ability to survive upheavals. The movement toward increasing gender diversity and racial diversity in senior management positions not only improves social equity, but also benefits corporations by introducing new perspectives, skill sets, and problem-solving styles.

The contrasts between living systems and engineered systems reflect a profound difference in their design. The theory of natural selection suggests that adaptability has been a key long-term success factor for living systems. Under stable ecosystem conditions, specialized life-forms were able to flourish, but when those conditions changed, only the most adaptable of species survived (including both humans and cockroaches). Engineered systems, however, are typically designed for precise operation under stable conditions so that their performance can be "optimized." Indeed, it would seem inefficient to design an engineered system to support a wide variety of operating conditions.

Of course, the solution to this dilemma is to combine the resilience of humans with the efficiency and precision of engineered systems. This happy marriage is the basis of the modern industrial enterprise, in which human organizations make key decisions that govern the deployment of vast amounts of capital, including manufacturing and transportation equipment. As mentioned in chapter 1, however, large enterprises may become more vulnerable to disruptions as they strive for greater efficiency, standardize their business processes, and become insulated from external changes. They begin to resemble machines more than living systems and may lose their capacity to sense and respond to threats.

It is no surprise, then, that the business community is showing a fresh interest in resilience. In the face of ever-increasing complexity, connectivity, and turbulence, it is time to abandon the mechanistic view of the enterprise as a controllable artifact and view it instead as a living system embedded in a dynamic network.

Operating in the Zone

Being "in the zone" or "in the groove" suggests a state of heightened performance, commonly seen in the world of sports. Scientists have long been fascinated with the ability of humans to demonstrate unusual skill by focusing intensely on a task and becoming oblivious to distractions. Some psychologists describe this state as "flow," a euphoric state of total immersion where the ego falls away, peak performance becomes effortless, and time seems suspended.[6] This state is familiar to athletes, musicians, and even some engineers. It also appears to be a universal characteristic of biological organisms.

A striking feature of living systems is the seemingly effortless way in which they sustain dynamic equilibrium. The remarkable resilience of biological organisms and ecosystems is inherent in their physical structures and behavior patterns, including the flows of materials, energy, and information. The subtle mechanisms that enable the resilience of living systems are only partially understood by science. Two of the keys to resilience in nature are structural variation and functional exploration.

1. Biological creatures exhibit two types of **structural variation**: at the organism level and at the species level. Because no two organisms are exactly alike, the natural world is full of variety, and selective advantages can emerge. Species evolution occurs on a longer time scale through the mechanisms of mutation and natural selection, enabling species to survive, adapt, transform, and flourish in response to fundamental changes in the environment.

2. Every organism is in rhythmic motion, interacting with its environment. There are many types of rhythmic motion, ranging from breathing to annual migration. In essence, natural systems are constantly experimenting and probing the world around them to maintain a balanced, dynamic equilibrium. **Functional exploration** is a continuous process, enabling living systems to sense fluctuations in their environment and respond accordingly.

Exploratory behavior enables living organisms to be constantly in the zone, intuitively making the correct choices to maximize their performance in pursuit of well-being. Some behaviors may be learned whereas others are genetically encoded, but the result is that whether hunting, mating, or migrating, living organisms will take appropriate actions without conscious analysis or forethought. There is no strategic planning in nature.

We exhibit the same types of autonomous behavior in our personal lives. We often rely on intuition and instinct, and our physical senses are attuned to the slightest stimuli so that even a pinprick will make us jump. We also have extraordinary capabilities for analysis, reflection, foresight, and planning—traits that are not found in the natural world. Although these analytic capabilities are helpful for understanding and managing complex systems, we should not allow ourselves to become detached from messy reality. The real world is infinitely more complex than the models and equations we use to describe it, so healthy skepticism is appropriate. In today's economy, equilibrium is fleeting, and knowledge of past and present patterns is insufficient to predict the future. However, many of our engineering and management practices are based on a static, linear worldview because individuals tend to cling to the familiar status quo.[7]

There are important lessons here for enterprise managers. Resilient organizations will typically have two modes of operation. First is analysis, planning, and design, during which the organization uses all available information to prepare for the opportunities and vulnerabilities that lie ahead. Second is moment-to-moment execution, during which the organization is applying its knowledge and skills to sense change and respond effectively. By learning from nature, managers should encourage a diversity of views and management styles during the planning mode and ensure that the organization stays alert during execution, sensitive to changes in the business environment and poised to respond effectively.

Dynamics of Enterprise Systems

Any business can be viewed as a living system whose performance is influenced both by the physical and intellectual aspects of its parts—people, processes, and assets—and by its network of relationships. Enterprises and natural systems follow similar patterns except that human foresight and intervention enable more rapid adaptation than nature does. In theory, the power of foresight should make humans more resilient, but we may not use that power effectively; our collective failure to take action with regard to climate change is a notable example. The source of both our power and our vulnerability is technology, and sometimes our clever solutions can have unexpected adverse consequences.

By viewing the enterprise as a system, resilience can be seen as the capacity to absorb disturbances and reorganize, retaining essentially the same function, structure, identity, and feedbacks.[8] Expanding on the definition in chapter 3, we can think of enterprise resilience in terms of dynamic behavior and define it as follows:

> A resilient enterprise continues to grow and evolve so as to meet the needs and expectations of its shareholders and stakeholders. It adapts successfully to disruptive changes by anticipating risks, recognizing opportunities, and designing robust products and processes.

Potential energy

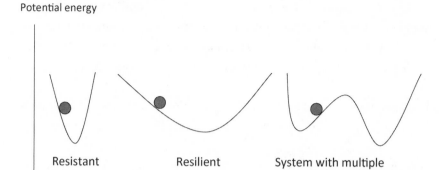

Adjacent system states

Figure 4.1. Examples of dynamic system behavior

By the laws of thermodynamics, closed systems will gradually decay from order into chaos, tending toward maximum entropy. Enterprise systems, like other living systems, are "open" in the sense that they continually draw on external sources of energy and maintain a stable state of low entropy that is far from thermodynamic equilibrium.[9] Perhaps the essence of sustainability is resilience, the ability to resist disorder.[10] Figure 4.1 provides a simplified illustration of thermodynamic changes that characterize different types of resilience. Imagine that the system state corresponds to a ball rolling along a curved surface. Each system has a stable state representing the lowest potential energy at which it maintains order, and each is subject to perturbations that shift it along a trajectory of adjacent states.

- The graph on the left in figure 4.1 is typical of engineered, highly controlled systems. These systems operate within a narrow band of possible states and are designed to *resist* perturbations from its equilibrium state. They recover rapidly from small perturbations, but may not survive a large perturbation.

- The middle graph in figure 4.1 is typical of social and ecological systems. These systems can function across a broad spectrum of possible states and tend to return gradually to their equilibrium

state. Through adaptation and evolution, they are capable of surviving large perturbations and thus are more resilient than a resistant system.

- The graph on the right in figure 4.1 is even more resilient than the
middle one in that this type of system can tolerate larger perturbations. Under certain conditions, the system may shift to a different
equilibrium state, representing a fundamental change in its structure
or function.

To ensure the survival and growth of an enterprise, managers must
understand how its performance is influenced by the changing patterns
in external systems to which it is coupled. Nowhere is this imperative
more evident than in the world of supply chain management (see chapter 6). The trends toward globalization and outsourcing have created
complex supply networks that are vulnerable to many types of disruptions. Economic volatility and international security concerns have only
increased the likelihood of such disruptions. In the automotive industry,
for example, supply chain executives discovered that the adoption of
"lean" production systems, which are highly efficient in a stable environment, increased susceptibility to business interruptions caused by schedule delays and other fluctuations.

During the 1990s, Royal Dutch Shell conducted a historical study of
corporations in an effort to understand what drives longevity. It found that
the average life expectancy of large corporations worldwide was less than
50 years; in effect, most companies die prematurely. Shell identified four
factors that distinguished longer-lived companies: sensitivity and adaptability to the business environment, cohesion and sense of identity, tolerance of diversity and decentralization, and conservative use of capital.[11]
Profitability was conspicuously absent from this list and was considered to
be an outcome rather than a predictor of longevity; many companies have
delivered spectacular profits for short periods and then vanished abruptly.
From this study emerged the notion that a corporation is best understood
as a resilient, adaptive organism rather than as a machine engineered to
deliver profits (table 4.1). In fact, resilience is essential for creating sustained, long-term profitability.

Similarly, the Evergreen Project, a decade-long study of 160 companies,
found that the main determinants of superior financial performance were

Table 4.1. Contrasting views of an enterprise

Profit Machine	Living System
• Owned by shareholders	• Guided by stakeholders
• Fulfills intended purpose	• Inherent purpose and identity
• Controllable, static	• Influenceable, evolving
• Designed and built	• Self-created and self-organized
• Responds to decisions	• Autonomous behavior
• Requires maintenance	• Regenerative capacity
• Uses human resources	• Human community
• Learns via employees	• Learns as an entity

not technology-based but rather reflected organic traits: an achievement-oriented culture, a flexible and responsive structure, a clear and focused strategy, and flawless execution.[12] This organic view of the corporation is consistent with an emerging recognition by the business community of intangible value drivers. Company characteristics such as customer trust, employee satisfaction, supply chain relationships, and corporate responsibility reputation are increasingly recognized as leading indicators of shareholder value (see chapter 5).

Despite these findings, most large enterprises are slow to respond to the challenges of turbulent change, as pointed out in chapter 1. Business processes tend to be mechanistic and rigid, focusing on repeatability rather than adaptation. Although it is clear that industrial systems and their environments are dynamic and nonlinear, decision makers tend to rely on linear, static models. For example, quality improvement methods seek predictability through standardization and efficiency through minimization of waste and redundancy. Business continuity planning is limited in that it tends to focus on company assets rather than external networks. The implication is that industrial systems will be robust only within their intended operating conditions and thus can be vulnerable to unanticipated fluctuations or simple human errors, as has been illustrated repeatedly. The 2003 crash of the space shuttle *Columbia*, the 2003 blackout of the Northeast electrical grid (see chapter 8), and the 2010 Gulf of Mexico oil spill are all examples of catastrophic failures in complex systems due to small, unforeseen disruptions.

The Adaptive Cycle

To understand enterprise resilience, it is helpful to examine how the concept of resilience has been addressed in different fields, including psychology, medicine, ecology, economics, and urban affairs. For example, psychologists define human resilience as the ability to transform adversity into a growth experience.[13] Although this analogy is helpful, enterprises are more complex than individuals and are more akin to ecosystems. Ecological resilience has been studied extensively by an international group of researchers led by Lance Gunderson and C. S. (Fritz) Holling. They have developed a general theory of adaptive cycles, arguing that all complex, adaptive systems exhibit similar patterns of slow accumulation of resources, increasing connectedness, and decreasing resilience, punctuated by periods of crisis, transformation, and renewal.[14] For example, mature forests are periodically destroyed by fire or vermin and then regenerate. Based on an understanding of these patterns, humans may be able to intervene in appropriate ways that take advantage of the system dynamics rather than merely resisting change.

In the business world, the adaptive cycle (figure 4.2) applies at many different levels, from the life cycle of a product to fluctuations in the global economy. The front loop of the adaptive cycle is similar to the well-known *S* curve, or logistic curve, which rises steeply and then flattens out due to

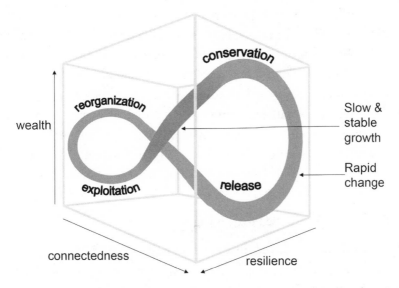

Figure 4.2. *The adaptive cycle in complex systems*

conservation of assets and constraints on growth. As wealth accumulates, the system stabilizes, with enterprises becoming more structured and less resilient than before. Eventually, the system is disrupted by any number of forces—industrial accidents, political upheavals, economic crises, disease epidemics, or technological failures—which leads to a collapse of the existing equilibrium and a release of accumulated assets. The system now enters a period of chaotic change and reorganization, corresponding to the back loop of the adaptive cycle, during which wealth is depleted and existing structures are fragmented. This process of creative destruction provides opportunities for exploitation of available assets and fresh innovation—new scientific discoveries, new institutions, new relationships, and new business processes—so that the system shifts into a more resilient state and reenters the growth phase.[15]

From the study of living systems, some basic principles have emerged that are applicable to understanding and enhancing the resilience of human systems, including both companies and communities (see "Resilience Principles for Living Systems" box). The adaptive cycle concept suggests that enterprises can prolong their growth phase and avoid collapse by continuously refreshing their knowledge, staying attuned to disruptive forces, and adapting to change. Even in the absence of traumatic disruptions, a "sense and respond" strategy increases awareness of change and enables agile implementation of midcourse corrections. Thus, we do not need to wait for long-term evolution to determine the survival of the fittest enterprises. Instead, through self-awareness and self-transformation, enterprises can redesign their structure and function to maximize their fitness for the journey ahead. For example, IBM has repeatedly reinvented itself to keep pace with the rapid evolution of information technology and services. In contrast, despite ample evidence of change, Polaroid was unable to make a successful transition from film to digital technology.

Resilience Principles for Living Systems

- **Resilience is an intrinsic characteristic of all living systems.** Living systems are purposeful, complex, adaptive, and self-organizing. They operate at many different scales, ranging from individual cells, to higher organisms, to sophisticated communities, to entire ecosystems.

- **Resilient systems exhibit awareness of and response to disruptions.** A living system is able to sense gradual disturbances or sudden threats and to respond via behavioral, functional, or structural adaptations that enable it to persist and preserve its identity (e.g., "fight or flight").

- **The evolution of living systems is influenced by cycles of change at multiple scales.** Every system is coupled with subsystems (e.g., components), higher-order systems (e.g., environments), and related systems (e.g., competitors). The associated cycles of change may be fast (e.g., flooding) or slow (e.g., global warming).

- **Resilient systems typically have corrective feedback loops to maintain a dynamic equilibrium.** Disruptions (e.g., invasive species) can shift a system away from equilibrium or can cause it to collapse. In response to disruptions, a system may cross a threshold and undergo a "regime shift" that leads to a different equilibrium state.

- **Self-organizing, self-aware systems can design for inherent resilience.** Human-designed systems (e.g., cities or enterprises) can learn to identify potential disruptions and to design their assets so that they can better absorb extreme events (e.g., graceful degradation) and adapt to a changing environment.

Global Challenges Ahead

The 3V framework introduced in chapter 3 explains how industrial supply chains and human communities use ecosystem services to create value while generating waste and emissions that flow back into the environment. Figure 4.3 shows some of the critical linkages among economic, environmental, and social systems that enable these systems to function in a resilient manner. These linkages include reliance on ecosystems for fulfillment of human needs as well as development of "shared value"

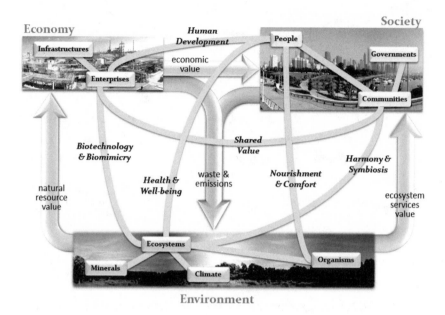

Figure 4.3. *Interdependencies among resilient systems*

between communities and enterprises (see chapter 5). To achieve sustainability, we must protect critical natural capital, improve resource productivity, and avoid environmental pollution, but unexpected disruptions can impair our ability to pursue this vision. To achieve resilience, we must encourage diversity, robustness, and adaptability in both natural and human resources as well as in governing institutions and supporting infrastructures.

Table 4.2 illustrates parallels in structural components and functional performance in three types of living systems: communities, enterprises, and managed ecosystems. To design for resilience, managers of these systems must modify either their functional processes or their structural configurations. These changes can range from short-term tactical adjustments to fundamental strategic transformations. The available methods for designing resilient systems are described further in parts 2 and 3 of this book.

According to the National Intelligence Council, it is estimated that by 2030 the world will need between 30 percent and 50 percent more water, energy, and food than it does today to keep up with rising demand.[16] In addition, we will need to provide those additional resources in ways that significantly reduce carbon dioxide emissions. Addressing any of these

Table 4.2. Examples of structure and function in living systems

System Type	Structural Components	Functional Performance
Urban community	Built environment, infrastructures, and commercial, residential, or other occupants	Provide goods and services to support occupants' economic and social activities
Enterprise supply chain	Network of assets, suppliers, manufacturers, logistics providers, and customers	Fulfill customer demand through physical, informational, and financial transactions
Cattle-grazing rangeland	Organisms (cattle, vegetation) and vital resources (air, soil, water, sunlight)	Nourish and sustain the web of interdependent living organisms

Figure 4.4. *Approximate resource flows across the energy-water-materials nexus in the United States*

resource needs individually is an immense task, but the challenge of ensuring sufficient supplies of water, energy, food, and other materials is magnified because of their interdependence. This so-called nexus is a source of great concern to both business strategists and environmentalists.

Figure 4.4 shows the interdependencies among energy, water, and materials in terms of resource flows; for example, about 26 gallons (100 liters) of water and 3 kilowatt-hours of energy on average are required to produce $1 worth of materials.[17] Similarly, about 550 lb (0.5 metric ton, or MT) of materials on average are required to generate 1 kilowatt-hour of energy. Because the potential effects of climate change will influence the cost and availability of all three resources, they must be managed intelligently and holistically so as to meet our resource future needs. In particular, it is important to avoid cascading failures by decoupling the infrastructures that supply basic commodities to cities and businesses.

It is clear that the vitality of our economy, environment, and society are closely intertwined and that issues such as energy, nutrition, health, and security cannot be neatly divided and conquered as separate problems. The sad truth, however, is that we are trying to address these complex, interdependent issues with limited data, outmoded tools, and a fragmented regulatory framework that was established decades ago. As Albert Einstein said, "We can't solve problems by using the same kind of thinking we used when we created them."

The resilience of the world economy was severely tested by the 2008 recession. In different but related ways, the resilience of the natural environment has been tested over the last one hundred years by the pressures of the expanding population and economy, including not just greenhouse gas emissions but also threats to biodiversity and resources such as air, water, soil, forests, wetlands, and minerals vital to the well-being of humans and other species. Ironically, the recession moderated these pressures, but as the global economic engine regains momentum, ensuring worldwide environmental resilience will be a formidable challenge. To address it, we need a different game plan for this century.

Takeaway Points

- Living systems, including individual organisms and colonies, have developed inherent resilience traits that enable them to cope with unexpected threats in a complex and changing environment.
- Operating "in the zone" is a state of effortless peak performance, often experienced by athletes, and seems to be an intrinsic behavior of living systems that contributes to their inherent resilience.

- Human enterprises should be viewed as living systems rather than complex machines, and they exhibit the same resilience principles and behavior patterns seen in natural systems.

- Both enterprises and natural systems go through a dynamic cycle of growth, maturity, collapse, and renewal, but human-managed systems can adapt more rapidly than nature and avoid severe disruptions.

- A systems view is necessary to understand interdependencies among human and natural systems; for example, the energy-water-materials "nexus" creates vulnerability for these critical resource flows.

- The challenges of the twenty-first century demand a holistic, integrated approach to the resilience and sustainability of enterprises as well as the social and natural capital upon which they depend.

PART 2

Practicing Enterprise Resilience

CHAPTER FIVE

Generating Business Value

Any company that can make sense of its environment, gener-
ate strategic options, and re-align its resources faster than its
rivals will enjoy a decisive advantage. This is the essence of
resilience.

Gary Hamel and Liisa Välikangas[1]

T
he case for enterprise resilience is clear: corporations need to
improve their capacity to survive, adapt, and flourish in the face of
unexpected disruptions. Resilience, however, is a relatively new concept
that is still being explored and digested by mainstream business leaders.
Putting it into practice will be a challenge for large organizations with
well-established business processes.

One possible approach is to practice resilience at a tactical level and
view it simply as an extension of existing processes such as security, emergency preparedness, business continuity planning, and risk management.
Many consultants and practitioners have already begun to rebrand their
traditional tools and techniques in this way. Unfortunately, this interpretation misses the important lessons of embracing change and systems
thinking, and is unlikely to produce competitive advantage. There is a
danger that "resilience" will become the latest buzzword, without any
deeper insights.

A more effective approach is for leaders to understand that resilience
requires a truly different mind-set, leading to new strategies for adaptation and growth. It may require fundamental rethinking and redesign of
existing business processes, leading to greater awareness of change and

heightened responsiveness to the inevitable stresses and shocks of the global business environment. It may also require closer engagement with customers, business partners, government agencies, and other stakeholders throughout the value chain.

An example of a chief executive who understood resilience as a strategic imperative is Peter Voser, former CEO of Royal Dutch Shell. Among major oil companies, Shell has displayed an unusual commitment to anticipating future trends and taking a proactive stance. Since the 1990s, Shell has done groundbreaking work on development of future scenarios as an input to strategic planning and has become a leader in implementation of corporate responsibility and sustainability practices.

Following the economic recession of 2008, Voser became concerned about the loss of trust between society and industry and wanted to reaffirm the importance of the private sector in enabling social progress. In 2010, he initiated a new phase of strategic thinking to broaden Shell's understanding of the emerging forces of change, including stresses such as the energy-water-food nexus discussed in chapter 4. Shell formed a senior-level team to explore the key factors that could make companies, cities, and nations resilient in the face of these stresses and to build partnerships that would enable systemic improvements in resilience.

Recognizing the importance of partnerships, Voser convened a group of like-minded chief executives from different industry sectors to discuss a collaborative approach toward resilience. Their first meeting was held in Davos, Switzerland, in early 2012 and resulted in the creation of the Resilience Action Initiative (RAI), an informal alliance with a pragmatic mission. They set out to work with cities, nongovernmental organizations, and selected academics to develop a better understanding of resilience and demonstrate how the resulting insights could be put into practice.

The membership of RAI eventually grew to include a diverse group of multinational companies: Dow, DuPont, Rio Tinto, McKinsey & Co., IBM, Unilever, Shell, Siemens, Swiss Re, and Yara. Their initial results, published in 2014, summarize their collective view:

> Large companies are part of society and need to be connected with its long-term requirements. So the focus of RAI is not only on the companies themselves, but also on the resilience of the cities and

regions in which they operate and where they eventually sell their products and services. The companies have found that by engaging with the resilience of their environment and that of their clients, they also strengthen their own resilience.[2]

The RAI partners have collaborated on numerous pilot projects around the world to explore resilience practices. In the United States, Shell is working on several projects with the City of Houston, the site of its North American headquarters. These projects include capturing waste materials to provide alternative sources of energy and brainstorming new approaches to urban mobility.

The Value Proposition

Resilience thinking is not strictly about avoiding setbacks. A key difference between resilience and traditional risk management is the potential for competitive advantage in a turbulent business environment. Companies that hone their resilience skills will be rewarded through better recognition of upside opportunities and better agility in capturing those opportunities. Rather than simply bouncing back, they bounce forward.

Several examples of seizing advantage in the wake of disruptions were identified by Yossi Sheffi in his pioneering analysis of enterprise resilience. For instance, the Los Angeles Metrolink transit system increased its ridership by a factor of twenty immediately following the January 1994 Northridge earthquake; FedEx seized an opportunity to fill unmet demand in the aftermath of a 1997 strike at UPS; and Dell took advantage of the West Coast port lockout in 2002 to spur demand for liquid crystal display monitors that could be shipped economically via air freight, displacing bulkier cathode-ray-tube monitors.[3]

As shown in figure 5.1, the value proposition for enterprise resilience can be understood readily in terms of shareholder value creation. There are two principal pathways—direct and indirect—for generating value.

The first pathway, **direct value creation**, occurs when resilience has a positive influence on economic returns, measured in terms of "tangible" benefits. To quantify such benefits, many chief financial officers use an economic value added (EVA) formula:

$$EVA = \text{after-tax operating profit} - \text{capital charge}$$

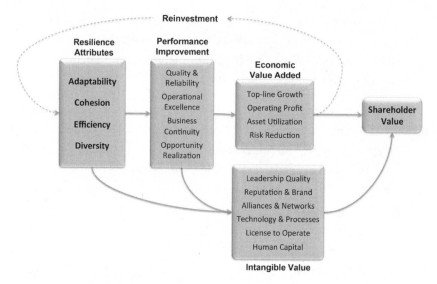

Figure 5.1. *How resilience contributes to shareholder value*

In essence, additional economic value can be generated either by increasing cash flow or by reducing the capital costs required to generate cash flow. Conversely, economic value may be lost by a decrease in cash flow or an increase in capital costs. Note that the bottom line, corresponding to the profit and loss statement, is just one part of the EVA equation. The other part is concerned with corporate assets and liabilities as reflected in the balance sheet. The capital charge is calculated by multiplying total asset value by weighted average cost of capital, which includes financing costs and risk premiums. Figure 5.1 shows the four main levers by which resilience can increase EVA:

1. **Top-line growth:** Resilient companies can increase revenues and expand market share through
   a. Differentiation from competitors based on quality and reliability of products and services
   b. Improvement of goodwill and customer loyalty in existing markets
   c. Capturing new market opportunities more rapidly than competitors
   d. Product and service innovation to meet the challenges of turbulence and uncertainty

2. **Operating profit:** Resilient companies can increase after-tax operating profits through

   a. Reduction in operating and maintenance costs based on productivity and operational efficiency

   b. Establishment of buffers and reserve capacity to avoid supply shortages and shipment delays

   c. Affordable access to critical resources such as energy, water, materials, and human talent

   d. Enabling reductions in insurance premiums or exclusions

3. **Asset utilization:** Resilient companies can reduce the complexity and fixed costs of assets through

   a. Process simplification and flexibility to adapt to shifting demand and changing conditions

   b. Improvement in reliability and availability of manufacturing and supply chain assets

   c. Assurance of business continuity without excessive redundancy or underutilized assets

4. **Risk reduction:** Resilient companies can reduce insurance costs and financial exposure through

   a. Prevention of unplanned disruptions that may lead to business interruption or liabilities

   b. Preparedness for crises to minimize downtime and mitigate losses or recovery costs

   c. Sustainable environmental management practices to avoid liabilities and regulatory delays

   d. Cultivation of a resourceful and collaborative employee culture to cope with unexpected events

The second pathway, **indirect value creation**, occurs when resilience has a positive influence on shareholder confidence, stakeholder trust, and company reputation. It is estimated that between 50 percent and 90 percent of a company's market value can be explained by "intangible" assets such as leadership, brand equity, and human capital rather than traditional measures such as earnings and financial assets.[4] Intangible assets include people, relationships, skills, and ideas that are not traditionally accounted for on the balance sheet.

Although economic returns provide a retrospective, or "lagging," indicator of value creation, it is important to consider these intangible, nonfinancial value drivers because they provide "leading" indicators of shareholder value improvement. To portfolio managers and investment analysts, intangible strengths are often the hidden clues that differentiate companies with comparable financial statements. In other words, improvements in resilience can strengthen a company's intangible assets in ways that lead to sustained long-term shareholder value.

Intangible Value Drivers

Intangible value is often difficult to monetize, but it is possible to assess and compare the strength of intangibles across a variety of industries. Based on considerable research, the following characteristics, depicted in figure 5.1, have been identified as among the most important intangible value drivers:[5]

- **Leadership quality:** Management capabilities, experience, vision for the future, transparency, accountability, and trust

- **Reputation and brand:** How the company is viewed globally in terms of stakeholder concerns, inclusion in "most admired company" lists, and sustainability performance; and strength of market position, ability to expand the market, perception of product/service quality, and investor confidence

- **Alliances and networks:** Customer and supply chain relationships and strategic alliances and partnerships

- **Technology and processes:** Strategy execution, information technology, inventory management, flexibility, quality, and internal transparency

- **License to operate:** Regulatory positioning, relationships with local communities, and ability to expand operations

- **Human capital:** Talent acquisition, workforce retention, employee relations, compensation, and perception as a "great place to work"

Companies can strengthen their intangible assets by improving resilience, both within and beyond their operations, in a number of ways. For example, they can assure continuity in the supply of critical resources for communities (e.g., energy, water, food), provide technologies and services to help communities recover from natural or other disasters, collaborate with communities to develop new approaches for infrastructure resilience, or deploy first responders to assist with disaster recovery and provide critical supplies.

The best of all worlds is when corporations simultaneously create economic value for their shareholders and improve the lives of their stakeholders, a concept that Michael Porter and Mark Kramer have described as "shared value."[6] As major investors demand improved corporate governance, transparency, and disclosure, shared value has become increasingly important. In addition to customers, shareholders, and employees, a broader collection of key stakeholders is concerned about economic and social resilience. They include suppliers and business partners; government officials at the local, state, and federal levels; neighboring communities; religious groups, advocacy groups, and other nongovernmental organizations; academic and research organizations; and, of course, the media. By responding to the concerns of these stakeholders, companies can strengthen their key relationships, reputation, and license to operate.

An example of shared value is Nestlé's work with coffee growers in impoverished rural areas of Africa and Latin America. The company provides them advice on farming practices, helps them secure bank loans, assists them in procurement of supplies, and provides them with cash incentives for higher-quality coffee beans. These practices not only benefit the growers; they also improve yields and reduce adverse environmental impacts. Most important for Nestlé is that they ensure the company a reliable supply of high-quality coffee, thus improving the resilience of its supply chain.

Table 5.1 provides examples of how the companies profiled in this book, as well as other resilient companies, have realized the benefits listed above. Of course, resilience does not always come free of charge. The value of improved resilience may be offset by expenditures on equipment and supplies, personnel training and preparedness, or other costs associated with business continuity and adaptation. Managers need to analyze

Table 5.1. Examples of resilience benefits in various industries

	Top-Line Growth	Operating Profit	Asset Utilization	Risk Reduction	Intangible Value
Dow Chemical (supply chain)	Expanding into new markets	Avoiding unnecessary costs	Right-sizing fixed assets	Improving business continuity	Gaining customer loyalty
Dow Chemical (environmental)		Harnessing ecosystem services	Reducing capital requirements		Protecting natural capital
IBM (supply chain)		Avoiding business interruptions		Rapid risk identification	
IBM (community)	Stimulating new markets for technology				Enabling smarter, more resilient cities
L Brands		Anticipating shipment delays		Avoiding business interruptions	
Royal Dutch Shell			Efficient use of waste and residuals		Community mobility and energy security
American Electric Power	Maintaining customer loyalty	Maintaining electrical service			Community recovery from power outages
Entergy			Planning deployment of critical assets	Adaptation on Gulf Coast to climate change	
Cisco	Designing resilient products	Avoiding business interruptions		Maintaining supply chain visibility	
DuPont	Products for personal safety under stress				Serving human needs
Veolia		Understanding total costs of water resources	Optimizing infrastructure capital costs	Reducing impact of water shortages	Improving brand value and license to operate

the costs and benefits of alternative approaches and develop a business case for resilience improvement. (Tools for measuring and managing resilience are discussed in chapter 9.) Finally, resilience may or may not align with other drivers of shareholder value, including short-term profitability and long-term enterprise sustainability, so it is important to weigh all these trade-offs when formulating a preferred strategy.

Four Resilience Attributes

By now, it should be clear that resilience is a universal concept, touching every aspect of an enterprise from physical facilities to human behaviors. To realize the strategic value proposition described above, each business unit will need to examine both its structural configurations and functional processes carefully. What should they be looking for? What makes one business more resilient than the next?

To decipher the drivers of enterprise resilience, it is helpful to think in terms of four systemic attributes, known by the acronym ACED and shown in figure 5.2: adaptability, cohesion, efficiency, and diversity.[7] Each attribute represents a cluster of resilience capabilities that are applicable to any business enterprise; definitions and examples are listed in table 5.2. We cannot simply try to maximize each of these attributes because some fundamental tensions need to be balanced.

Adaptability and efficiency are fundamentally opposed because pursuit of efficiency will tend to eliminate sources of variability and unused

Figure 5.2. Fundamental attributes of enterprise resilience

Table 5.2. Fundamental enterprise resilience attributes

	Definition	Examples
Adaptability	Capacity to adjust or transform in response to changing conditions	• Flexibility in procurement and logistics • Capacity to handle surges in demand • Ability to recover from natural disasters
Cohesion	Existence of unifying forces or linkages that preserve continuity	• Strong corporate or brand identity • Employee loyalty and teamwork • Strategic partnerships with suppliers
Efficiency	Capacity to operate effectively with modest resource consumption	• Streamlined operating procedures • Ability to decouple from utility grids • Recovery and reuse of waste materials
Diversity	Existence of multiple talents and styles that enable innovation	• Cultural and ethnic diversity • Diversification of business lines • Tolerance of different business models

capacity. As mentioned in chapter 4, this tension has emerged in the realm of supply chain operations, where the push toward lean strategies has increased the vulnerability of supply chains to unexpected disruptions. The current thinking in operational excellence involves a compromise approach that is both lean and agile, or "leagile."

Likewise, cohesion and diversity are fundamentally opposed because pursuit of cohesion will tend to eliminate diversity of talents, opinions, and business models. Organizational studies show that as enterprises grow larger, they become more rigid and hierarchical, tend to resist change, and become vulnerable to unexpected disruptions (see chapter 8). On the other hand, although smaller organizations tend to be loosely structured and more creative, they are susceptible to anarchy and dissent. The challenge of creating a resilient corporate culture is to encourage individuality and resourcefulness while maintaining a sense of common identity and purposeful teamwork.

The implication of these tensions is that resilience requires a balance between opposing attributes. To achieve balanced resilience, an enterprise must find the right mix of resilience capabilities that fit with the characteristics of its businesses, while addressing the vulnerabilities that each business may face. Chapter 6 illustrates this strategic approach in the context of supply chain resilience and provides a finer-grain representation of the four fundamental attributes.

Resilience Attributes

- **Adaptability** can be defined as the capacity of a business to adjust or transform in response to changing conditions. Doing so may involve modifying the business's structure or function due to either actual changes or anticipated shifts in prices, market demand, competitive pressures, or other trends in the business environment. It may also involve responses to perceived threats, ranging from natural disasters to deliberate sabotage. The time frame over which adaptation occurs can vary tremendously, from minutes to decades. Some authors prefer to distinguish between *agility* as a response to sudden shocks versus adaptability as evolution over time.[8] Others consider adaptability as the capacity to preserve the existing system versus *transformability* as the capacity to fundamentally change the system.[9] The many organizational capabilities associated with adaptability, including anticipation, flexibility, and recovery, are described further in chapter 6, which deals with supply chain resilience.

- **Cohesion** can be defined as the existence of unifying forces or linkages that preserve the continuity of a business. A distinctive corporate culture and corporate identity, supported by strong values and principles, are examples of such forces. Employee pride and loyalty are the hallmarks of cohesive enterprises, and their employees are likely to make personal sacrifices and take unusual action beyond the norm to support the company in times of crisis. Cohesive companies are often characterized by strong, visionary leaders who inspire employees and other stakeholders. They demonstrate *stability* through good times and bad and are able to achieve *longevity* if they are sufficiently adaptable.

- **Efficiency** can be defined as the capacity to operate effectively with modest resource consumption. Companies that use resources carefully and judiciously are able to increase their *productivity* and tend to perform better than other companies in the event of resource scarcity or price increases. In contrast, companies that are habitually wasteful may be challenged to adjust to unexpected shortages of energy or raw materials and may have difficulty in surviving economic downturns. The lean movement has emphasized driving waste out of the system to increase *profitability*. There is, however, a danger in carrying efficiency too far, to the point where it erodes buffers and redundancies such as safety stock. Cost optimization methods are often based on steady-state assumptions and may not account for the increasing volatility of the business environment.

- **Diversity** can be defined as the existence of multiple talents and styles that enable innovation in response to changing conditions. There are many forms of diversity, focusing on race, gender, culture, profession, or personal characteristics. Apart from questions of *social equity*, combining a diverse set of skills and perspectives generally improves creative problem solving and *resourcefulness* in response to unexpected disruptions. If different groups of employees within a company form factions, however, overall company cohesion and collaborative teamwork may suffer.

The dynamic interplay among these resilience attributes is evident in the concept of the innovator's dilemma, which argues that large, mature companies are inevitably displaced by the external emergence of disruptive technologies.[10] As companies grow and become established in their markets, they initially gain resilience by growing more efficient and more cohesive. Over time, however, they may lose diversity due to uniformity of established thinking and may lose adaptability due to size and inertia. In a stable environment, such companies may survive for long periods, but they will typically be vulnerable to disruptions that stem not only from technological innovation, but also from changes in any of the systems to which they are linked. According to the adaptive cycle theory described in chapter 4, they will eventually reach a point of rapid decline where their old business model collapses, wealth is redistributed, and new businesses emerge to fill the vacuum. By encouraging strategic diversity and organizational agility, established companies may be able to forestall their decline. Some long-lived companies like 3M and IBM have demonstrated that they can avoid enterprise sclerosis and sustain a resilient culture, continually reinventing their business models in response to a changing environment. Like resilient biological species, these companies focus on assuring the adaptability and diversity of their genetic essence—their DNA—rather than on preserving an existing structure.

Resilience in Business Processes

Adopting enterprise resilience does not imply creating a new depart- ment or treating resilience as yet another business function. Resilience thinking should be seamlessly embedded into every business process. Each of the four ACED attributes can be leveraged to improve the resilience of specific business processes, as summarized in table 5.3 and as follows:

- In **supply chain management**, operational resilience is improved by streamlining and developing flexible alternatives; strategic resil- ience is increased by encouraging innovation and establishing infor- mation connectivity. Chapter 6 addresses supply chain resilience in greater detail.

- In **environmental, health, and safety (EHS) management**, proac- tive efforts to develop new technologies for conserving resources and eliminating wastes can help reduce costs, avoid regulatory burdens, and develop trust among key constituencies. For example, in 2002, Baxter Healthcare's annual environmental financial statement showed a savings of $65 million, about three times its annual expenditures. At the same time, implementation of enterprise-wide EHS management systems and effective crisis management helped minimize the effect of disruptions and ensured business continuity. Chapter 7 addresses environmental resilience in greater detail.

- In **human resource management**, employee satisfaction and growth are key prerequisites for enterprise productivity and continuous learn- ing. For example, a Watson Wyatt study of 405 public companies found that a well-managed workforce can add up to 30 percent to a company's market value.[11] However, in some cases resilience driv- ers may conflict; for example, strengthening cultural cohesion may reduce the diversity of perspectives and business models. Chapter 8 addresses organizational resilience in greater detail.

Table 5.3. Strategies for enhancing enterprise resilience

	Adaptability	Cohesion	Efficiency	Diversity
Supply chain management	Develop alternate supply channels and flexible options	Improve visibility via information connectivity	Reduce inventory needs; improve response time	Encourage supplier creativity and innovation
Environmental, health, and safety management	Enhance capability for crisis response and failure recovery	Develop enterprise-wide environmental, health, and safety management policies and principles	Encourage closed-loop recovery and reuse of materials	Explore alternative sources of raw materials and energy
Human resource management	Track key trends and encourage adaptive learning	Develop strong corporate culture and identity	Raise productivity of labor; reduce absenteeism	Encourage diverse perspectives and business models
Product and service development	Reduce cycle time for innovation and deployment	Develop strong brand identity and reputation	Increase ratio of value delivered to total cost of ownership	Develop multiple configurations and extensions
Capital asset management	Develop flexible, multifunctional equipment	Coordinate asset deployment throughout the enterprise	Improve asset availability and utilization ratio	Invest in emerging and disruptive technologies
Customer relationship management	Anticipate and detect important market shifts	Develop strategic alliances with key customers	Facilitate and expedite customer support	Enable flexible customization of offerings
External affairs management	Explore and prepare for external change scenarios	Emphasize social responsibility and citizenship	Leverage external resources and partnerships	Maintain dialogue with stakeholder groups

- In **product and service development**, strategic resilience is achieved through an innovation process that emphasizes value creation for all stakeholders: increasing the ratio of customer value to total cost of ownership, reducing the cycle time for development and deployment (thus increasing market share and profitability), enabling product adaptation through multiple configurations and extensions, and strengthening brand identity and company reputation. Companies like General Electric, IBM, and Siemens are already betting that radical innovations in ubiquitous digital technology (commonly known as the Internet of Things) will displace established product and service technologies. When the displacement begins, perhaps triggered by a slight shift in economic factors, we may see a rapid restructuring of industries ranging from energy production to retailing. Chapter 10 describes how designing for resilience can be incorporated into the innovation process.

- In **capital asset management**, resilience is increased by improving asset utilization to achieve leaner operations, developing flexible resources, diversifying the portfolio of available technologies, and coordinating asset deployment across the enterprise. For example, Anheuser-Busch reengineered its North American transportation and warehousing network to cope with increasing product complexity, achieving a reduction in transportation costs of 15 percent while cutting wholesaler out-of-stock time by 30 percent.

- In **information technology management**, resilience is achieved by providing redundancies and backup systems to maintain essential communications, avoid loss of valuable data, and prevent unauthorized intrusion. Hackers have developed increasingly sophisticated cyber-crime techniques, ranging from malware schemes to theft of intellectual property, and internal fraud and abuse are common threats. In February 2014, the Obama administration launched a cybersecurity framework, providing guidelines to help industry and government organizations strengthen the security and resilience of critical infrastructure.

- In **customer relationship management**, resilience means staying close to the customer via efficient support services and strategic alliances, sensing emerging changes in the market, and being flexible in tailoring customer solutions. For example, since 1990, Ashland has operated a total chemical management business in support of its electronics industry customers, enabling chip manufacturers to outsource full responsibility for chemical acquisition, storage, quality assurance, waste disposition, and EHS compliance.

- Finally, in **external affairs management**, resilience is increased by leveraging external resources and partnerships, exploring and preparing for external change scenarios, maintaining open dialogue with diverse stakeholder groups, and emphasizing social responsibility and corporate citizenship. For example, FedEx Express has enhanced its brand by building a reputation for environmental leadership, partnering with external groups such as the Environmental Defense Fund, and adopting sustainable technologies such as 100 percent recycled packaging and hybrid diesel-powered delivery vans.

For each of the above business processes, measures of adaptability, cohesion, efficiency, and diversity can complement traditional performance measures because they are *leading* indicators, reflecting fundamental attributes of a business that drive both profitability and sustainability. Moreover, resilience offers a hierarchical approach for linking macroscale systems with finer scales of resolution; for example, a business might assess the resilience of specific product systems and then aggregate these assessments to the enterprise level. It is also possible to combine various resilience indicators into an *enterprise resilience index*. Doing so will allow companies to benchmark their overall performance and to identify leaders and laggards in resilience among their facilities, products, processes, or business units. Chapter 6 describes the development of the SCRAM resilience index for supply chain management.

The Chief Resilience Officer

Given the complementarity between enterprise risk management and resilience, it makes sense for the chief risk officer (CRO) to take on the added responsibility of maintaining and enhancing enterprise resilience. The scope of resilience is a good deal broader than risk management, however, and the CRO's job description and skill set will need to expand considerably. Rather than simply "playing defense" against undesired liabilities and business disruptions, the CRO will need to consider strategies and opportunities for creating competitive advantage by adapting to changing conditions. It is plausible that *chief resilience officer* positions with the added dimension of business value creation will become commonplace in the coming years.

The Rockefeller Foundation established a bold new precedent as part of its 100 Resilient Cities challenge program, which aims to help cities around the world survive, adapt, and grow no matter what kinds of chronic stresses and acute shocks they experience. Urban resilience is not just about responding to disasters; it is also about dealing with stresses such as unemployment, violence, and food or water shortages. In this ongoing program, selected cities are awarded about $1 million each, and each receives financial and logistical guidance as well as access to expertise for developing its resilience strategies. As part of the program, each city appoints a chief resilience officer to oversee these efforts, and their experiences will most likely influence similar developments in the private sector.

The following suggests the responsibilities that might be assigned to the chief resilience officer of a business enterprise:

- Work with senior management and the board of directors to establish a comprehensive vision of enterprise resilience as a key requirement for sustainable competitive advantage.

- Deploy established enterprise risk management processes to identify, evaluate, and control known threats and risks that may degrade or interrupt business operations.

- Engage with business and functional leaders to determine how resilience thinking can complement existing risk management processes and deliver competitive advantage.

- Develop or acquire tools to assess the vulnerability and resilience of company assets and business processes, and apply these tools to evaluate the resilience of all major business operations as well as key collaborators, customers, and supply chain partners.

- Develop and implement effective resilience strategies that leverage the strengths of the company and improve both structural and functional resilience at a reasonable cost.

- Develop appropriate indicators of risk and resilience, establish continuous improvement goals, and report on enterprise progress toward those goals.

- Provide education, training, and communication materials to inform employees and stakeholders about the enterprise's commitment to risk management and resilience.

- Engage with thought leaders in business, government, and academia to promote policies and practices that improve the resilience of external systems that are important to the enterprise, including communities, natural resources, and critical infrastructures.

Successful performance in the CRO role will require a range of personal qualities, including vision, leadership, and positive management style; the ability to communicate complex concepts to a variety of audiences; the ability to build effective teams and relationships, both internally and externally; familiarity with risk management and resilience theory and practice; familiarity with a broad range of corporate functions and operations; and the ability to plan and manage multiple simultaneous projects and initiatives.

Takeaway Points

- Enterprise resilience is emerging as a strategic imperative in an age of turbulence. The formation of the Resilience Action Initiative is an early indicator of corporate interest.

- The value proposition for resilience includes direct benefits in terms of economic value added as well as indirect benefits in terms of strategic positioning and stakeholder satisfaction.

- The fundamental attributes of a resilient enterprise are adaptability, cohesion, efficiency, and diversity, but these attributes involve fundamental tensions that must be balanced.

- Awareness of resilience should be incorporated into existing business processes, including supply chain management, environmental management, human resource management, and innovation management.

- The chief risk officer should morph into a broader role as chief resilience officer to ensure that resilience is seamlessly incorporated into the company's strategic planning, decision making, and external engagement processes.

CHAPTER SIX

Resilience in Supply Chain Management[1]

A supply chain is only as strong as its weakest link.

Anonymous

Contemporary supply chain managers face a variety of global forces that increase the potential for unforeseen disruptions. Of all the enterprise functions, supply chain management is the most vulnerable to external stresses and shocks. Supply chain disruptions—even minor shipment delays—can cause significant financial losses for companies and substantially affect shareholder value.

A recent report published by the World Economic Forum surveys the current landscape of global supply chain disruptions.[2] Respondents to the forum's survey ranked natural disasters, sudden demand shocks, extreme weather conditions, information and communication disruptions, and political unrest as the most important risks that can affect supply chains (figure 6.1). Their responses indicate that some of the most important threats to supply chains today are neither preventable nor under the direct control of management.

Although the indirect consequences of disruptions are often difficult to quantify, changes in stock price can be an indicator of these effects. That same report analyzed how stock markets responded in the aftermath of several major global disruptions (figure 6.2). Hurricane Katrina caused only a modest decline in the S&P 500, but other events caused steep drops in value.[3] Previous research has shown that announcements of supply

Environmental	Natural disasters	59%	
	Extreme weather	30%	
	Pandemic	11%	
Geopolitical	Conflict and political unrest	46%	
	Export/import restrictions	33%	
	Terrorism	32%	
	Corruption	17%	
	Illicit trade and organized crime	15%	
	Maritime piracy	9%	
	Nuclear/biological/chemical weapons	6%	
Economic	Sudden demand shocks	44%	
	Extreme volatility in commodity prices	30%	
	Border delays	26%	
	Currency fluctuations	26%	
	Global energy shortages	19%	
	Ownership/investment restrictions	17%	
	Shortage of labour	17%	
Technological	Information and communications disruptions	30%	
	Transport infrastructure failures	6%	

■ Uncontrollable
▨ Influenceable
■ Controllable

Source: World Economic Forum Supply Chain and Transport Risk Survey 2011

Figure 6.1. *Relative importance of supply chain disruption triggers*

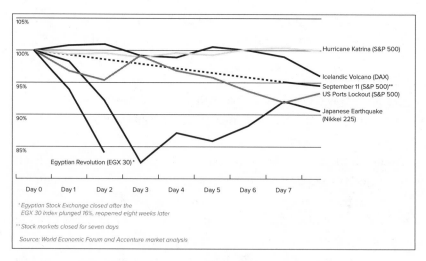

Figure 6.2. *Stock market responses to global events*

chain problems are correlated with abnormal decreases in shareholder value of about 10 percent.[4]

The increasing intensity and frequency of natural disasters is perhaps the most visible factor affecting supply chain performance. For example, in the wake of Superstorm Sandy, seaports located in the northeastern United States were closed to containerized traffic, while many freight carriers, including CSX, Norfolk Southern, and YRC Freight, either suspended their services or warned customers to expect delays.[5] To supply its delivery trucks, FedEx rented fuel tankers because commercial gas stations ran dry.[6]

A notable example of supply chain resilience to disasters is the response of DHL to the 2010 eruption of the Eyjafjallajökull volcano in Iceland, which grounded millions of air cargo shipments for several months. DHL activated its emergency plan and rescheduled one hundred flights from its hub in Leipzig, Germany, toward destinations in southern Europe that did not experience airspace closures. Furthermore, it quickly shifted transport to ground vehicles by deploying a fleet of trucks to Leipzig to retrieve shipments, while providing continuous status updates to its customers. As a result, DHL not only avoided significant financial impacts but also took the opportunity to reevaluate the most cost-effective combination of road and air transport.[7]

Although enterprises tend to focus on the supply side of their supply chains when scanning for potential risk factors, they also need to pay attention to the customer side. Increasing volatility and sudden changes in demand are important factors that can affect a firm's operations and ultimately its revenue. For example, in March 2013, Cardinal Health announced that its pharmaceutical distribution contract with Walgreen Co. would not be renewed after August 2013 because Walgreen decided to switch to another pharmaceutical distributor. Walgreen was one of Cardinal Health's largest customers, accounting for about 21 percent of revenue for 2012, and Cardinal Health's share price fell by 8.2 percent immediately after the announcement.[8] The company was able to recover quickly and continue its growth, however, thanks to deliberate efforts to expand and diversify its customer base. Upside shocks can be equally disruptive; companies that experience a sudden surge in demand may lack the capacity to fulfill orders and may lose potential business as a consequence.

Supply Chain Resilience Challenges

Supply chain practices that work well in a stable business environment may no longer be viable, and may in fact reduce competitiveness. In particular, just-in-time production and lean management seek to reduce inventories, minimize waste and process variability, and control information exchange tightly. These practices are often achieved by developing close relationships with a small number of suppliers. They do, however, make supply chains vulnerable to disruptions due to the lack of reserve capacity.

Many manufacturing companies have sought to balance their lean approach with "agile" practices.[9] Some argue that supply chains should develop structural flexibility, defined as "the ability to adapt to fundamental changes in the business environment."[10] An example of a structurally flexible organization is Zara, which has developed a rapid-fire supply chain that is extremely responsive to the ever-changing fashion environment and to customers' preferences. To achieve this type of resilience, it may be necessary to sacrifice some efficiency.

Due to the globalization of trade, supply chains have become longer and more complex, thus increasing the likelihood of disruptions. Moreover, anticipating disruptions in advance and managing them when they do occur have become extremely challenging. Because complex global supply networks are characterized by limited visibility, potential risks are hidden, and their potential cascading effects may not be understood. An often-cited example is Nokia's cell phone business, which discovered in 2000 that one of its key suppliers in New Mexico was concealing the fact that its facility had been destroyed by fire. Early recognition of the crisis enabled Nokia to secure alternative supplies, and it modified the product design to broaden its sourcing options, ultimately gaining significant market share. In contrast, Nokia's competitor, Ericsson, which relied on the same supplier, lost about $400 million in sales due to slowness in crisis recognition and response, and eventually exited the cell phone business.[11] Ironically, in subsequent years, Nokia stumbled in terms of its strategic resilience, failing to compete successfully in the smartphone market.

Globalization was initially driven by companies locating their plants in countries with lower labor costs and less stringent regulations. Disruptions such as the 2010 volcanic eruption in Iceland and the 2011 tsunami

Table 6.1. Examples of reshoring initiatives

Company	Operations
Caterpillar	Shifted some production from Japan to a manufacturing site near Athens, Georgia, to build small tractors and excavators
Chesapeake Bay Candle	Shifted the production of its candles and home-fragrance products from China and Vietnam to Maryland
General Electric	Moved the manufacturing of washing machines, refrigerators, and heaters from China to Louisville, Kentucky
Google	Built production capacity for its Nexus Q in San Jose, California
NCR	Opened a plant in Columbus, Georgia, where it builds automated teller machines and self-service checkout systems, moving operations from China, Hungary and Brazil

Source: B. McMeekin and E. McMackin, "Reshoring U.S. Manufacturing: A Wave of the Present," white paper, September 2012; and "Reshoring Manufacturing: Coming Home," *The Economist*, January 2013.

in Japan, however, revealed the vulnerabilities of extended supply chains. For example, 41 percent of manufacturers surveyed by the Federal Reserve Bank of Minneapolis indicated that the tsunami in Japan had affected them negatively.[12]

In recent years, many manufacturers have reevaluated their sourcing options and are considering reshoring (also known as backshoring or onshoring); that is, they are considering shifting operations back to their home markets. Table 6.1 lists examples of companies that plan to or have already reshored significant portions of their manufacturing operations. Although these companies had many motivations for reshoring, including improved customer service and stimulation of the domestic economy, reducing their exposure to risk was also an important driver.

Sustainability Pressures

A variety of forces have led companies to adopt a commitment to sustainability, also known as corporate environmental and social responsibility. These forces range from governmental regulations and stakeholder pressures to a desire for differentiation and innovative leadership. In a 2013 survey of more than one thousand CEOs, about half stated that sustainability will be very important to the future success of their business, but 67 percent acknowledge that the business community is not doing enough to meet sustainability challenges such as poverty, climate change, water scarcity, ecosystem degradation, and mineral depletion.[13]

There is increasing public awareness and concern about safe and ethical practices in consumer product supply chains. For example, in November 2012, it was revealed that IKEA had tolerated the use of forced prison labor during the Cold War by its suppliers in communist East Germany. Similarly, in August 2011, Zara was accused of accepting dire worker conditions and use of slave labor by its suppliers in Brazil. There have been numerous instances of mass casualties due to fires in offshore textile factories. Some industries, such as electronics and apparel, have formed worldwide coalitions that are demanding supplier compliance to strict codes of environmental and social responsibility.

Another global trend that increases the vulnerability of supply chains is climate change, which already appears to be causing rising sea levels and greater volatility in weather patterns. One consequence of climate change is severe droughts, creating water scarcity and threatening crops and livestock, across the world, including in the United States. Many companies have recognized that extreme precipitation and flooding as well as extreme droughts can increase their supply chain risks.[14] For example, because water and natural ingredients are critical components of its consumer products, Johnson & Johnson is developing strategies to address water scarcity. Indeed, water stewardship is a concern for a wide spectrum of manufacturing sectors as well as the energy sector. The oil and gas industries inject millions of gallons of water annually into aging oil fields to improve production, and electric power plants account for about half of annual water withdrawals in the United States.[15]

The explosion of social networking and digital media has transformed the business environment, creating greater transparency but also potential for abuse. Due to these new forms of communication, information travels instantaneously, reaches more people, and can persist longer on the Internet. Moreover, receivers of information often have no way of judging its accuracy or credibility, and the adverse effects of misinformation on brand image can be substantial. Viral dissemination of supply chain incidents or failures can damage company goodwill and license to operate; for example, a fake BP public relations Twitter account emerged after the BP oil spill in 2010 in the Gulf of Mexico, and attracted many more followers than the genuine BP feed.[16] Thus, social media pose extreme challenges to companies, including lack of visibility regarding information dissemination and lack of remedies to control the damage.

Balancing Vulnerabilities and Capabilities

Global enterprises need to cultivate supply chain resilience by recognizing their vulnerabilities and developing specific capabilities to cope with disruptions. As we saw in chapter 4, companies can try to emulate some of the behaviors seen in natural systems, such as tolerance for variability, continuous adaptation, and exploitation of opportunities created by disruptive forces. Resilient supply chains do not fail in the face of disturbances; rather, they adapt and maintain their ability to deliver products to the customer.

A team of researchers from the Colleges of Engineering and Business at Ohio State University, working with several companies, including fashion retailer L Brands Inc. (formerly known as Limited Brands), Dow Chemical, Johnson & Johnson, and Unilever, developed a comprehensive framework for supply chain resilience assessment and management (SCRAM).[17] The team generated an exhaustive catalog of vulnerabilities and capabilities based on existing literature as well as interviews and focus groups with managers and employees of companies that had experienced supply chain disruptions (table 6.2). Subsequently, the team identified linkages between specific vulnerabilities and capabilities, enabling the identification of proactive strategies for resilience improvement.

SCRAM identifies six major types of supply chain **vulnerabilities** (table 6.3), defined as "factors that make an enterprise susceptible to disruptions."[18] One frequently cited factor is **turbulence**. In the context of SCRAM, turbulence is defined as changes in the business environment that are beyond a company's control, including shifts in customer demand, geopolitical disruptions, natural disasters, and pandemics. Another category of vulnerability is **deliberate threats**, such as theft, sabotage, terrorism, and disputes with labor or other groups. Additional

Table 6.2. Supply chain vulnerabilities and capabilities

	Definition	Factors in SCRAM Framework	Fundamental Grouping
Supply chain vulnerabilities	Attributes that make an enterprise susceptible to disruptions	Turbulence	
		Deliberate threats	
		External pressures	
		Resource limits	
		Sensitivity	
		Connectivity	
Supply chain capabilities	Attributes that enable an enterprise to anticipate and overcome disruptions	Flexibility in sourcing	Adaptability
		Flexibility in manufacturing	Adaptability
		Flexibility in order fulfillment	Adaptability
		Capacity	Adaptability
		Adaptation	Adaptability
		Anticipation	Adaptability
		Recovery	Adaptability
		Dispersion	Diversity
		Operating efficiency	Efficiency
		Visibility	Efficiency
		Collaboration	Cohesion
		Organization	Cohesion, diversity
		Market position	Cohesion
		Security	Cohesion
		Financial strength	Cohesion
		Product stewardship	Cohesion

vulnerabilities come from **external pressures** that create constraints or barriers (such as innovations, regulatory changes and shifts in cultural attitudes); **resource limits** that have the potential to constrain a company's capacity (such as availability of raw materials or skilled workers); the **sensitivity** and the complexity of the production process; and the degree of **connectivity** in the company's supply chain, implying a need for coordination with outside partners.

Similarly, SCRAM identifies sixteen **capabilities** that companies can deploy to respond to their particular vulnerability patterns (table 6.4). Capabilities are defined as "factors that enable an enterprise to anticipate and overcome disruptions."[19] These factors are classified in table 6.2 according to the fundamental enterprise resilience attributes defined in chapter 5; the emphasis is clearly on adaptability and cohesion.

Table 6.3. Supply chain vulnerability factors

Vulnerability Factor	Definition	Subfactors
Turbulence	Environment characterized by frequent changes in external factors beyond the firm's control	Unpredictability in demand, fluctuations in currencies and prices, geopolitical disruptions, natural disasters, technology failures, pandemics
Deliberate threats	Intentional attacks aimed at disrupting operations or causing human or financial harm	Terrorism and sabotage, piracy and theft, union activities, special-interest groups, industrial espionage, product liability
External pressures	Influences, not specifically targeting the firm, that create business constraints or barriers	Competitive innovation, government regulations, price pressures, corporate responsibility, social/cultural issues, environmental, health, and safety concerns
Resource limits	Constraints on output based on availability of the factors of production	Raw material availability, utilities availability, human resources, natural resources
Sensitivity	Importance of carefully controlled conditions for product and process integrity	Restricted materials, supply purity, stringency of manufacturing, fragility of handling, complexity of operations, reliability of equipment, safety hazards, visibility of disruption to stakeholders, symbolic profile of brand, customer requirements for quality
Connectivity	Degree of interdependence and reliance on outside entities	Scale and extent of supply network, import/export channels, reliance on specialty sources, reliance on information flow, degree of outsourcing

Note: A firm is indirectly vulnerable to disruptions that affect their multiple tiers of customers and suppliers. The SCRAM framework can be used to assess the resilience of selected external organizations.

Table 6.4. Supply chain capability factors

Capability	Definition	Subfactors
Flexibility in sourcing	Ability to quickly change inputs or the mode of receiving inputs	Common product platforms, supply contract flexibility, supplier capacity, supplier expediting, alternate suppliers
Flexibility in manufacturing	Ability to change the quantity and type of outputs quickly and efficiently	Product/service modularity, multiple pathways and skills, manufacturing postponement, change-over speed, batch size, manufacturing expediting, reconfigurability, scalability, rerouting of requirements
Flexibility in order fulfillment	Ability to change the method of delivering outputs quickly	Multisourcing, demand pooling, inventory management, alternate distribution modes, multiple service centers, transportation capacity, transportation expediting
Capacity	Availability of assets to enable sustained production levels	Backup utilities, raw materials, reserve capacity, labor capacity, ecological capacity
Efficiency	Capability to produce outputs with minimum resource requirements	Labor productivity, asset utilization, quality management, preventive maintenance, process standardization, resource productivity
Visibility	Knowledge of the status of operating assets and the environment	Information technology, status of inventory/equipment/personnel, information exchange with supplies/customers/carriers, market visibility, external monitoring
Adaptability	Ability to modify operations in response to challenges or opportunities	Seizing advantage from disruptions, alternative technology development, learning from experience, strategic gaming and simulation, environmental sustainability

Capability	Definition	Subfactors
Anticipation	Ability to discern potential future events or situations	Demand forecasting methods, risk identification and prioritization, monitoring/communicating deviations and "near misses," recognition of early warning signals, business continuity planning, emergency preparedness, recognition of opportunities, business intelligence gathering, government lobbying, awareness of global change
Recovery	Ability to return to normal operational state rapidly	Equipment reparability, resource mobilization, communications strategy, crisis management, consequence mitigation
Dispersion	Broad distribution or decentralization of assets	Distributed suppliers/production/distribution, distributed decision making, location-specific empowerment, dispersion of markets
Collaboration	Ability to work effectively with other entities for mutual benefit	Collaborative forecasting, supply chain communication, collaborative decision making, supplier/customer involvement in innovation, postponement of orders, product life cycle management, supplier/customer collaboration, risk/reward sharing with partners
Organization	Human resource structures, policies, skills, and culture	Creative problem-solving culture, accountability, diversity of skills and experience, substitute leadership capacity, benchmarking/feedback, culture of caring for employees, workforce flexibility

(Continued)

Table 6.4. (Continued)

Capability	Definition	Subfactors
Market position	Status of a company or its products in specific markets	Brand equity, customer loyalty/retention, market share, product differentiation, sustainability position
Security	Defense against deliberate intrusion or attack	Layered defenses, access restriction, employee involvement in security, collaboration with governments, cybersecurity, personnel security
Financial strength	Capacity to absorb fluctuations in cash flow	Financial reserves and liquidity, portfolio diversification, insurance coverage, price margin
Product stewardship	Sustainable business practices throughout the product life cycle	Proactive product design, resource conservation, auditing and monitoring, supplier management, customer support

Finally, SCRAM identifies 311 separate "linkages" whereby specific capabilities can counteract specific vulnerabilities. These linkages provide a methodology for identifying the most important vulnerabilities and setting priorities for capabilities that need to be strengthened.

For example, a company that faces unpredictable market demand could improve flexibility by strengthening several capabilities: increased manufacturing capacity to satisfy surges in demand; accurate, up-to-date visibility of demand status to support timely decision making; early recognition of market changes to enable strategic responses; and close collaboration with customers and suppliers to ensure coordinated action. Similarly, a company concerned with dependence on a complex supply network could improve reliability by identifying alternative sources, reducing lead times, and increasing its inventory buffers. Based on the results of the SCRAM analysis, managers can develop a portfolio of capabilities to address important resilience gaps and strengthen overall competitiveness.[20]

Although similar organizations are likely to share some similar features, different companies—and even different business units *within* companies—will have their own set of vulnerabilities and capabilities. SCRAM provides a qualitative measure of supply chain resilience, which increases as capabilities increase and decreases as vulnerabilities increase (figure 6.3). An organization with high vulnerabilities that does not have adequate capabilities will be overexposed to risks; in response, it should invest resources in improving the particular capabilities in question. Conversely, an organization that faces low vulnerabilities but invests heavily in capabilities may be eroding its profits unnecessarily. Thus, there is no "one size fits all" approach. Organizations need to find their zone of balanced resilience by developing the right portfolio of capabilities to fit the pattern of vulnerabilities they face.

The SCRAM approach has been fully adopted at Dow Chemical, which has implemented it in more than twenty of its global business units and has achieved significant business benefits through its use. For example, Dow applied the SCRAM process to its P-Series Glycol Ethers family of products. Dow SCRAM team members identified several disruption scenarios for further analysis: production site shutdown, raw material supply outage, and internal raw material allocation shortage. They developed a simulation model to test the consequences of these scenarios and were able to confirm a 95 percent service level with their existing capabilities. This analysis led to right-sizing fixed assets and working capital,

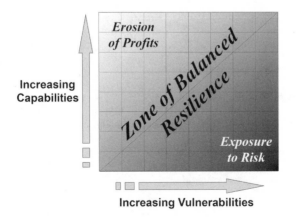

Figure 6.3. *The zone of balanced resilience*

representing a $1.1 million savings for this business and a 500 percent return on modeling effort. The Dow team was recognized as a finalist in an innovation award competition by the Council for Supply Chain Management Professionals.[21]

The above approach represents a systems view of supply chain dynamics, helping companies understand the inherent vulnerabilities that could lead to disruptions and the capabilities that are within their control. The systems view is illustrated in figure 6.4, where each arrow represents a causal linkage, either an amplifying effect (+) or a dampening effect (−). Reducing vulnerabilities may be possible (broken-line arrow), and by increasing capabilities in a cost effective manner, companies can improve their supply chain performance. Unexpected disruptions can adversely affect supply chain performance by increasing the cost to serve customers and reducing customer satisfaction. By learning from experience and developing a better understanding of their vulnerabilities and capabilities, companies can reduce the frequency of disruptions and the severity of their impacts, thereby generating both increased customer satisfaction and reduced supply chain operating costs. Although reducing external

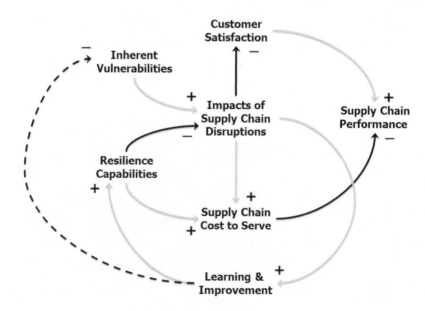

Figure 6.4. System dynamics of supply chain resilience

pressures and turbulence may be difficult, there are many options for improving capabilities. The cost of such improvements must be balanced against the expected supply chain performance benefits.

Resilience in Action

Recovering from the 2011 Japanese Tsunami: A Tale of Three Companies

General Motors (GM) spends about 2 percent of its purchasing budget in Japan. The 2011 tsunami there revealed several vulnerabilities, including exposure to natural disasters, supplier dependencies, and supply availability. For example, GM was forced to shut down its Chevrolet Colorado and GMC Canyon plant in Shreveport, Louisiana, temporarily because it lacked components that were supplied from Japan. GM's Japanese counterparts struggled for two months to resume production, but GM was able to recover rapidly because, four days after the earthquake, GM mobilized its resources and initiated a disaster recovery action plan called Project J. A team of GM employees gathered in three crisis rooms from which they monitored the disaster and its consequences. They used supply chain mapping tools to identify components sourced by Japanese suppliers, categorize the status of the suppliers, and map the affected product lines. This exercise helped to identify 118 problematic components. The team then took action by idling several plants to conserve supplies and seeking alternative suppliers for some of the parts. Further, GM employees visited the affected suppliers in Japan to assess the damages firsthand and help them recover. Through Project J, GM was able to resolve 113 of the problematic components and managed to avoid any material effect on its earnings.

Cisco became aware of the tsunami through alerts from a third-party notification service. Within twelve hours, Cisco's supply chain risk management team, composed of cross-functional managers, was activated and was able to identify all direct suppliers, their associated sites and components, and other critical nodes that were within the affected area. Further, the manager was able to profile each supplier from various perspectives, such as expected time to recover, backup generation capabilities, or whether the components were single sourced. Contact with suppliers was established within a few days to assess the ability of the supplier to continue to supply components. A snapshot of each supplier's status was developed and refreshed on a daily basis, facilitating rapid and informed decision making. In addition, the company established a war room within two days through which the team handled the decision making and communication of the team's strategy to the relevant stakeholders. Although Cisco's business continuity plans helped it respond to the disruption, it was the company's investments in resilience capabilities—such as improved information sharing, common information platforms, engagement with suppliers to anticipate outcomes, and constant communication with customers—that helped the company emerge from the disaster unscathed. Moreover, Cisco's partners praised the way the company handled this disaster, demonstrating increased customer satisfaction.

In contrast to GM and Cisco, Merck was adversely affected by the tsunami disaster due to lack of resilience. The company produced 100 percent of the global supply of Xirallic, a pearl-luster pigment for automotive paint, in its factory in northeastern Japan. Following the tsunami, Merck closed its plant for two months and aimed to have another production line ready in Germany by the end of that year to continue with the production of the pigment. Meanwhile, Merck did not allow its customers to enter its plant to assess damages and help with restoring operations; instead, it kept them in the dark. The factory closure displeased Merck's customers, including Toyota, Nissan Motor Co., Ford Motor Co., Chrysler, and Volkswagen AG; subsequently, several of them chose to develop the pigment in-house or search for other suppliers. Merck could have improved its resilience in several ways, such as dispersing production of Xirallic across several sites, improving flexibility by enabling rapid transfer of operations from Japan to Germany, improving collaboration with customers, and building anticipatory capabilities such as scenario planning and emergency preparedness.

Resilience in Action

Assuring Continuity at L Brands[22]

The twenty-nine ports on the West Coast of the United States handle most of the nation's containerized cargo import traffic from the Pacific Rim countries and account for about 12.5 percent of the nation's GDP.[23] A classic test of resilience occurred during the fall of 2002, when these ports were locked down by the Pacific Maritime Association (PMA) as a result of stalled negotiations with the International Longshore and Warehouse Union (ILWU).

PMA is responsible for negotiating and administering maritime labor agreements with the ILWU. During the contract negotiations in 2002, both sides insisted on introducing clauses that they viewed as important. PMA wanted to introduce new automation technology to expedite movement of cargo, something its members had been demanding unsuccessfully throughout the late 1990s, while the ILWU lobbied for an increase in its pension fund.

The potential for a strike became increasingly clear when the contract officially expired on July 1, 2002, and was not extended. Fearing the loss of jobs, the union refused to accept any technology implementation and began to use the tactic of slowdowns in work. Finally, in September 2002, PMA decided to lock the workers out, forcing a shutdown of all the ports. After government intervention, the ILWU and PMA resumed negotiations. They eventually agreed on the implementation of a computerized cargo tracking system and an increase in pension funding. In addition, the duration of the contract was doubled to six years.

The lockout paralyzed US commerce. Imports worth billions of dollars, ranging from basic commodities to electronic products, were stranded on huge ocean vessels that were too large to pass through the Panama Canal. Before long, there were more than one hundred vessels backed up around the giant port of Long Beach, California. Perishable foods, such as meat from Australia, bananas from Guatemala, and grains from Asia, were waiting to enter, while California grapes and apples were waiting to leave.

Compounding this problem was another import from Asia: the just-in-time management technique, which was implemented by many US companies during the 1990s. During the port lockout, auto companies such as Honda and Mitsubishi and some defense contractors had to delay production due to the lack of sufficient parts inventory. Retailers with minimal safety stock or seasonal demand feared excessive stock-outs during the upcoming Christmas season, which typically accounts for more than one third of their annual sales.

Although the contractual dispute causing the lockout was resolved within only ten days, many manufacturers and retailers suffered long-lasting consequences. Economists estimated a total loss of sales of up to $1 billion per day for US companies and a backlog of up to one hundred days for recovery of trapped cargo.

Tensions between PMA and the ILWU continue today. The ports typically approach 90 percent to 95 percent capacity during peak months, while the volume of ocean cargo keeps increasing each year. Productivity pressures have driven increased use of automation at the expense of longshoremen positions, which has only aggravated the situation. In 2008 and again in 2014, as the six-year labor contract renewal approached, the pattern was repeated. In 2014, the ILWU caused a slowdown in traffic that lasted nine months,; it was finally resolved in February 2015.

Resilience at L Brands

L Brands (formerly Limited Brands) sells women's and men's apparel, lingerie, and beauty and personal care products. In 2002, at the time of the lockout, it encompassed more than 3,700 stores, including Victoria's Secret, Bath & Body Works, Express, Henri Bendel, The Limited, and several other brands. The company's annual sales were around $11 billion, and it employed about 100,000 associates throughout the United States.

Due to the fast-changing and unpredictable nature of the fashion business, resilience is ingrained into the company's culture and is an integral part of the business continuity planning and enterprise risk management processes. L Brands ensures continuity of critical functions by monitoring the development of potential threats to the business while taking a balanced approach to the assumption of risks.

The responsibility for supply chain resilience rests with Lbrands Logistics Services (LLS), a wholly owned operating division that manages the global supply chain for the company. Mike Sherman, former senior vice president for transportation and logistics, described the LLS approach for managing business fluctuations as follows:

> It is part of our DNA to deal with risk and be proactive. For instance, we shifted cargo from the Port of Haifa to the Port of Ashdod in expectation of the tension between Israel and Lebanon. Unfortunately, we cannot predict every single scenario nor build total redundancy. We can only be proactive whenever possible, evaluate all the options, and make a fast decision on next steps.

Even though L Brands is primarily a specialty retailing company, LLS has turned logistics operations into a core competency. The company's size and scale have enabled the development of advanced systems for security and business continuity, comparable to those of leading retailers such as Target and Wal-Mart. These resilient capabilities provide competitive advantage over smaller-sized firms and reduce the risk of business disruptions.

A Strategic Approach to the West Coast Lockout

Prior to the 2002 West Coast port lockout, L Brands had experienced the PMA-ILWU tensions during the 1999 contract renewal. Anticipating the contract renegotiation in 2002, the company devised a contingency plan to address a possible slowdown in July when the contract was about to expire. LLS worked actively with the entire family of brands on risk mitigation strategies to prevent interruptions. The company gradually reduced its dependence on

West Coast ports by shipping ahead of schedule, rerouting products through the Panama Canal, and using airborne freight.

July came and went, however, and the LLS detailed plan for responding to a strike was never played out. The company was exposed to constant risk for the next three months, waiting to see how the ILWU-PMA negotiations would evolve. LLS instructed the brands to consider every shipment, decide whether they could accept a two-week delay, and explore alternative arrangements. LLS also held weekly conference calls with the brands to brief them on the situation.

After the lockout in October 2002, LLS moved to plan B, using alternative channels. Rick Jackson, former senior vice president at LLS at that time and now executive vice president for logistics at L Brands, explained how events played out:

> We had daily conference calls with our associates in Hong Kong, Long Beach, and Columbus and maintained 24-hour coverage during the lockout period. We put in a lot of resources and effort to manage through this experience, because protecting the business is a part of our culture. We were not thinking about the transportation budget, but the overall profit of the business.

Thanks to its defensive strategies, L Brands had almost no goods stranded on the ocean. Even though there were some nonmaterial cost increases in transportation, implementing those defensive measures avoided millions of dollars of potential losses, according to Jackson. In contrast, other retailers suffered significant financial effects.

Resilience Strategies at L Brands

The approach to resilience at L Brands during the 2002 West Coast port lockout is an example of a **sense and respond** strategy for coping with uncertainty about potential business disruptions (see chapter 1). Based on the SCRAM framework, the company demonstrates the following capabilities for assuring supply chain continuity:

Anticipation: To anticipate a potential crisis, LLS collected information from trade and general news; from network data for port capacity and volume; and from brokers, shippers, and on-site employees.

Adaptability: The company implemented a sustained monitoring and replanning process throughout the prolonged crisis, never taking the situation for granted.

Flexibility in sourcing: The company established alternate suppliers in other countries and designed its contracts to ensure flexibility in procurement of goods.

Flexibility in order fulfillment: The company identified surrogate distribution channels such as alternative ports and air freight opportunities for products.

Capacity: Early shipping and excess in-transit and safety stock inventory ensured extra capacity.

Recovery: The company had standard operating procedures for regular communication orchestrated by crisis response teams.

Market position: Strong market share and high volume allowed for charter airlifts where needed. Brand strength allowed for some back orders from catalog and delayed purchases at stores.

Financial strength: A diversified product line with high margins made it possible to absorb excess costs such as air shipment.

Visibility: High supply chain visibility, such as knowledge of product locations and estimated arrival dates, enabled the company to reroute shipments and order additional stock.

Continuous Vigilance

The West Coast port lockout provided L Brands with another in a series of experiences at handling major disruptions. The company successfully endured this test of resilience due to its well-developed culture of continuous vigilance. To facilitate rapid decision making, L Brands had already developed crisis management teams and communications protocols as a standard operating procedure following the terrorists attacks of September 11, 2001. Cross-functional teamwork and communication throughout the organization via frequent conference calls were an important success factor in coping with this prolonged crisis.

In addition, the company's risk management culture was a key factor in its ability to react to the unexpected events that emerged. By not fixating on a single scenario of how the situation would evolve, the company was prepared to face uncertainty and think holistically rather than narrowly. LLS protected the entire business; rather than focusing solely on transportation efficiency, it considered the overall profitability of the company. As a result, LLS gained credibility for its expertise in managing uncertainty, including its problem-solving attitude and capacity for quick response. The resulting increased confidence from the brands has paid back in similar dynamic situations that required unified follow-through, such as periodic labor and political disruptions and catastrophic environmental events in key sourcing regions.

From a business continuity perspective, L Brands continuously monitors the development of potential threats to the business. A major thrust, for example, is planning for a potential avian influenza pandemic, which could affect 40 percent of the labor force and disrupt vital sectors of the economy, as predicted by the World Health Organization.

To enable rapid response to any type of disruption, L Brands continues to improve its global intelligence gathering regarding political instability. The company is also working closely with suppliers on practicing for the unexpected, not just with labor strikes or pandemics, but with other scenarios, such as building fires or natural disasters.

To deal more effectively with emerging threats, the company continues to partner with industry leaders. For example,

L Brands is a part of the Waterfront Coalition, which promotes efficient and technologically advanced ports in the United States and abroad. Another major initiative across the company is an enterprise-wide "end-to-end" information technology integration called INSIGHT. As part of this integration effort, L Brands is addressing governance and contingency planning through a special cross-functional team.

These proactive approaches, including continuous vigilance, internal and external partnerships, system integration, and standard operating procedures for information sharing, are essential for anticipating and responding to threats. At L Brands, resilience goes beyond conventional business continuity and security: it is a pervasive mind-set that has been deliberately designed into the company culture.

Takeaway Points

- Enterprise supply chains are exposed to pressures such as climate variability and natural disasters, demand volatility, and political and economic fluctuations.

- Lean strategies such as just-in-time manufacturing are no longer viable, and supply chain managers are striving to increase agility and buffer capacity.

- Globalization of trade has created longer supply chains with decreased visibility, but many companies are reversing the off-shoring trend and restoring domestic operations.

- Social media have amplified sustainability pressures, including concerns over human rights, poverty, water scarcity, ecosystem degradation, mineral depletion, and ethical conduct.

- A comprehensive framework has been developed for supply chain resilience assessment and management (SCRAM) based on balancing of vulnerabilities and capabilities.

- A balanced approach to resilience will select the right portfolio of capabilities to offset the most important business vulnerabilities, avoiding excessive investments that might erode profits.

- **Resilience in Action:** Resilience was demonstrated in the way three global companies dealt with supply chain disruptions caused by the 2011 earthquake and tsunami that destroyed the Fukushima nuclear power plant in Japan.

- **Resilience in Action:** Fashion retailer L Brands has repeatedly demonstrated how resilience is ingrained into the company culture, illustrated by its adept handling of the West Coast port lockout.

CHAPTER SEVEN

Resilience in Environmental Management[1]

*You cannot tackle hunger, disease, and poverty unless you can
also provide people with a healthy ecosystem in which their
economies can grow.*

Gro Harlem Brundtland

In today's tightly coupled global economy, the definition of national
security is changing. Security is no longer merely concerned with
defense against hostile regimes and terrorist attacks; now it also includes
protection of our sources of food, energy, water, and materials, which
are the foundation of economic growth and community prosperity. In
2014, the US Department of Defense identified climate change as one of
the greatest immediate threats to national security. According to Defense
Secretary Chuck Hagel:

> Rising global temperatures, changing precipitation patterns, climb-
> ing sea levels, and more extreme weather events will intensify the
> challenges of global instability, hunger, poverty, and conflict. They
> will likely lead to food and water shortages, pandemic disease,
> disputes over refugees and resources, and destruction by natural
> disasters in regions across the globe. In our defense strategy, we
> refer to climate change as a "threat multiplier" because it has the
> potential to exacerbate many of the challenges we are dealing with
> today—from infectious disease to terrorism. We are already begin-
> ning to see some of these impacts.[2]

The linkage between environment and security has a long history, underscored by events such as the oil embargo of 1972 that led to gas rationing in the United States. That same year, a well-known report by the Club of Rome highlighted the "limits to growth" due to natural resource scarcities and continuing deterioration of environmental quality.[3] It also identified an array of socioeconomic problems such as urbanization and migration, particularly in developing countries, that could lead to security threats or violent conflicts. Indeed, there are several environmentally-related problems with potential security implications, including natural disasters, territorial disputes, population growth, and resource scarcity.

In 1987, the United Nations' Brundtland Commission report advanced the idea that "the whole notion of security as traditionally understood—in terms of political and national threats to sovereignty—must be expanded to include the growing impacts of environmental stress."[4] The report advanced the vision of an economically sounder and fairer future based on policies and behavior that can secure our ecological foundation. Today, the importance of ecosystem services to the health of the global economy has been recognized by the business community thanks to the leadership of various organizations, including the World Business Council for Sustainable Development (WBCSD).

Beginning in the 1990s, the linkage of environment and security began to appear in high-level US policy statements. The *National Security Strategy*, a document that states US foreign and security policy objectives, advanced the notion in 1991 that the United States should seek "cooperative international solutions to key environmental challenges, assuring the sustainability and environmental security of the planet as well as growth and opportunity for all."[5] Eleven years later, the 2002 *National Security Strategy* stated: "A world where some live in comfort and plenty, while half of the human race lives on less than $2 a day, is neither just nor stable. Including all the world's poor in an expanding circle of development— and opportunity—is a moral imperative and one of the top priorities of U.S. international policy."[6]

Meanwhile, EPA's Science Advisory Board observed that "competition for natural resources like ocean fish and potable water may pose as much of a threat to international political stability as an interrupted oil supply does today."[7] The board recommended that the United States develop strategic policies linking national security, foreign relations, environmental quality, and economic growth.

This broader view of national security reflects new global pressures that now threaten the well-being and resilience of both human society and the natural environment. These pressures include population growth, increased demand for energy and materials, and competition for access to land, water, minerals, and other vital natural resources. The resulting impacts include changes in global climate and degradation of clean air and water, soil, forests, and wetlands, all of which have the potential to compromise energy security, food security, supply chain security, and other domestic and international concerns.

The vitality of our planetary ecosystems is already seriously threatened. According to an international study led by the Stockholm Resilience Center, we have significantly exceeded the planetary safe operating boundaries in terms of greenhouse gas emissions, nitrogen flows, freshwater consumption, and biodiversity.[8] The 2005 Millennium Ecosystem Assessment found that fifteen of twenty-four important global ecosystem services are being degraded or used unsustainably.[9] Soon, global ecosystems will be under even greater pressure: by 2050, global population will increase by 30 percent to about nine billion people. Even if population growth slows, poverty alleviation and rising affluence in developing nations will inevitably increase the demand for natural resources and the generation of greenhouse gas emissions. As a consequence of these trends, climate and ecological disruptions could lead to widespread conflict over resources.[10]

Resolving these global challenges is the responsibility of national governments and international bodies such as the United Nations. No single enterprise can solve these problems on its own. Companies can lead by example, however, and can make a difference by working collectively through organizations such as WBCSD and Business for Social Responsibility. This chapter focuses on strategic environmental management opportunities. Later, chapter 12 addresses the broader question of how to balance the trade-offs between enterprise sustainability and resilience.

Environmental Threats to Enterprise Resilience

The 3V framework introduced in chapter 3 is helpful for understanding the importance of environmental issues for enterprise resilience. This systems view makes it clear that companies need to be conscious of their interdependencies with social and environmental systems. During normal operation, progressive companies will strive to achieve a sustainable, dynamic equilibrium in which the flows of resources are balanced with the health and vitality of social and environmental systems. If those systems are fragile or threatened, the resilience of the enterprise may be compromised. Figure 7.1 depicts some of the important dependencies that need to be considered.

Enterprises are dependent on natural resources both directly and indirectly. Most companies require land for siting their operating facilities. Primary manufacturing companies require raw materials, including minerals, water, biomass, fuels, and other commodities that are extracted from the environment. In addition, from a product life cycle perspective, companies are dependent on the availability of environmental resources to support their extended supply chains as discussed in chapter 6. Of particular importance is the continuous availability of utilities and infrastructure, which are heavily dependent on environmental

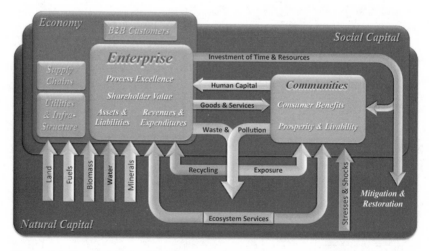

Figure 7.1. Systems view of environmental resilience: Resource flows and interdependencies

resources and are especially vulnerable to environmental disturbances such as earthquakes.

At the same time, enterprise operations entail a variety of environmental risks that can affect business continuity and impose additional costs or delays. They include risks of noncompliance with regulatory restrictions on environmental releases or discharges of waste and emissions, as well as requirements for occupational and public health and safety. Technological failures or human errors can lead to incidents such as chemical spills, pipeline leaks, or accidental process failures that can result in significant costs and loss of goodwill. Moreover, enterprises may be liable for environmental problems that were caused by other parties, including supplier or customer negligence and historic waste disposal practices.

Based on the SCRAM framework, the following vulnerabilities that are relevant to the environmental resilience of an enterprise and its stakeholders have been identified:

- **Natural disasters and pandemics.** Perhaps the greatest threat to business continuity is the occurrence of unexpected environmental disruptions such as hurricanes, tornadoes, droughts, earthquakes, volcanic eruptions, tsunamis, floods, and mudslides. As pointed out in chapter 1, the frequency and severity of such events appear to be increasing, and the lead times for defensive action are typically very short. Pandemics are also a concern, as evidenced by the Ebola epidemic in West Africa during 2014. Most companies have developed emergency plans to cope with these types of occurrences, but it is impossible to anticipate and protect against all the environmental threats associated with the entire network of customers, suppliers, contractors, municipalities, and other enterprise stakeholders.

- **Deliberate attacks.** Many companies are the target of attacks by interest groups who (perhaps unfairly) portray them as degrading the environment for the sake of profit. Despite the best efforts of companies to engage with stakeholders and demonstrate environmental responsibility, these types of attacks are inevitable, often unpredictable, and sometimes irrational. The attacks may come from several different stakeholder groups: employees who act as whistleblowers, activist shareholders, special interests such as religious and environmental groups, and competing firms. Attacks may arrive through a variety of mechanisms, including litigation, business obstruction, and news or social media.

- **Regulatory and market pressures.** Environmental awareness has stimulated the growth of markets for "greener" products that not only satisfy traditional customer needs and regulatory constraints, but also provide environmental benefits. The practice of "design for environment" has now been embraced by most major industries, ranging from primary producers of energy and materials to manufacturers of consumer goods such as automobiles, electronics, groceries, and home appliances.[11] To remain competitive, many companies have intensified their environmental efforts, and many have felt obliged to apply for eco-labels that certify their products and processes. At the same time, any evidence of environmental problems, such as contamination of products with trace carcinogenic substances, can significantly affect customer confidence and brand image.

- **Resource limits.** As shown in figure 7.1, natural resources such as land, fuels, biomass, minerals, and water provide feedstocks for energy generation, manufacturing, agriculture, and many other economic activities. Some of these provisioning resources are finite and are being rapidly depleted, including freshwater, fossil fuels, and rare earths. Other resources such as food crops are renewable, but the increasing demands of a growing global population exceed the capacity of ecosystems to regenerate these resources. In addition, unexpected disruptions or infrastructure breakdowns such as a bridge collapse can interrupt the supply of fuels or other critical resources.

- **Ecosystem services.** Apart from the provisioning services that supply natural resources, a variety of ecosystem services are fundamental to the economy yet are largely taken for granted. These services include absorbing pollutants to help preserve clean air and water; stabilizing the climate and water cycles and buffering against natural disasters; sequestering carbon to reduce global warming; maintaining soil fertility and preventing erosion; pest regulation; and pollination. Degradation or loss of these ecosystem services can impose significant costs to industry and consumers; for example, the colony collapse disorder that has afflicted honeybees in the United States placed billions of dollars' worth of food crops at risk.

In addition to these dependencies on natural capital, enterprises depend on social capital to provide economic stability; enforce the rule of law; and create productive, living communities that promote education and produce a skilled workforce. Community prosperity and quality of life, however, are also dependent on ecosystem services and are vulnerable to environmental disruptions. Therefore, improving enterprise resilience involves not just looking inward at company-owned assets, but also investing time and resources externally to ensure the resilience of communities where the enterprise manufactures or sells its products as well as the resilience of ecosystems that support both these communities and company operations.

The concept of *creating shared value*, as practiced by Nestlé and others, is a good example of how individual companies can be a positive force in advancing global environmental security (see chapter 5). In addition to cost reductions and productivity improvements, companies can raise the living standards of people around the world and thus contribute to a more sustainable and resilient society.

Companies can shore up their resilience by improving environmental systems and processes in many ways, ranging from closed-loop resource recovery (see chapter 4) to life cycle product stewardship. One advantage of taking a systems view is that companies can identify opportunities to leverage the resilient characteristics of ecological systems. The Dow Chemical example below describes the company's success in harnessing ecosystem services in the form of green infrastructure, thus providing industrial waste treatment capacity at a fraction of the traditional engineering cost. Chapter 10 describes many more opportunities for companies to design for resilience systematically.

Finally, there is an urgent need for enterprises to collaborate with national and regional governments to address global environmental vulnerabilities. Failure to achieve an effective response will lead to adverse outcomes that pose threats to national security and competitiveness, namely, conflicts over resources such as land, water, energy, and materials; lack of readiness for climate change impacts, leading to economic disruptions and community displacement; adverse health events, including spread of disease; and economic hardship that in turn threatens social stability.

Toward Global Environmental Security

The 3V framework in figure 7.1 highlights the linkages in industrial, societal, and environmental systems, showing that the quality and adequacy of environmental resources provide an essential foundation for both industrial value creation and societal well-being. A variety of international relationships—trade and tourism, foreign investment, mutual aid and alliances, and education and migration, for example—keep these dynamic systems in balance. National security involves assuring the smooth functioning of these relationships and avoiding disruptions due to natural or anthropogenic causes.

To respond effectively to planetary resource pressures and avoid national and international conflicts, we can look to six principal areas of response.

1. The practice of **risk assessment and management** is commonly used to set environmental regulations and protect human health. Similarly, risk management is used in security practices to analyze threats and countermeasures and to support effective long-range planning. Failure to anticipate risks can have severe consequences, as exemplified by the 2010 Gulf of Mexico oil spill. The National Commission on the BP Deepwater Horizon Spill and Offshore Drilling concluded that the risk assessments conducted by BP were insufficient and that government oversight was severely compromised.[12] The agency in charge of promoting the expansion of drilling—resulting in more than $18 billion in oil revenues—was also in charge of safety. As described in chapter 2, traditional risk management practices need to be supplemented by improved anticipation and awareness.

2. A second major response is exercising **control over resource consumption** from both land and oceans, including food, water energy, and materials. The consumption of staple crops such as wheat, rice, corn, and soybeans has outstripped production, and the once large stockpiles of these commodities have seriously declined. Future grain production is likely to be adversely affected by climate change and associated weather extremes. Similarly, logging and extraction of minerals such as coal, metals, and rare earths for industrial products must be approached with greater awareness of resource limits.

Our oceans have been overfished, and our freshwater supplies are seriously depleted. Careful land use is necessary to protect the vital ecosystems that support the US economy. Failure to control resource extraction can lead to conflicts over food and fuel use, water, and mineral rights.

3. Development of **urban infrastructure** is also a pressing need as more and more people move to urban areas. In the United States, aging infrastructures, including roads, bridges, and pipelines, have been neglected for years and pose public safety risks. In other parts of the world, development of new infrastructure is sorely needed. According to the United Nations, about 54 percent of the world's population currently lives in urban areas; this number is expected to increase to 66 percent by 2050, with most of the increase concentrated in Asia and Africa.[13] Such growth will create challenges not only for water, power, and transportation infrastructures, but also for protection of human health and safety.

4. The urgency of **poverty alleviation** is a key element of the UN Millennium Development Goals[14] and is closely tied to environmental security. We live in a world with 2.9 billion people surviving on less than \$2 a day, 2.6 billion people without access to proper sanitation, 1.2 billion people without access to safe drinking water, 924 million people dwelling in slums, 829 million people chronically undernourished, 790 million people lacking health services, and 39 million adults and children living with HIV / AIDS.[15] In addition, far more people are dying of malnutrition and disease than of conflict or war.

5. A critical need is **protection of human health** in developing nations. More than 100 million people around the world are estimated to be at risk from elevated levels of toxic pollutants, including lead, mercury, chromium, arsenic, pesticides, and radionuclides.[16] These risks to human health are largely a consequence of industrial activities, yet a thriving industrial base is essential for economic development and social well-being. This conflict could potentially be resolved through the introduction of innovative technologies, including design of safer and more environmentally conscious products using green chemistry and engineering.[17]

6. Finally, the need for **climate adaptation** has become evident. Gradual stresses such as sea-level rise and temperature change, combined with the increasing frequency and intensity of extreme weather events such as droughts and hurricanes, will force companies and communities to restructure their built environment and develop defensive measures. When planning investments in climate adaptation, managers should take advantage of the opportunity to think more broadly about the full spectrum of stresses, shocks, hazards, and risks that need to be considered as well as the full range of systems that would benefit from resilience enhancement (see chapter 11).

Taking a systems view that recognizes the convergence of environmental protection and security will enable a deeper understanding of the potential interactions among environmental drivers and strategic responses. It will not only lead to improved security, but will reinforce economic competitiveness for US industry and quality of life for US communities.

Resilience in Action

Ecosystem Services at Dow Chemical[18]

Dow Chemical was among the first US multinational companies to make a corporate commitment to sustainability and has been a consistent leader in sustainable product innovation and sustainability reporting. In recent years, Dow has begun to emphasize enterprise reliability as a key competitive factor. The company defines enterprise reliability as "the ability to constantly and consistently meet stakeholder expectations and commitments." Reliability includes assurance to customers that they will receive on-time delivery of high-quality products as well as assurance to other stakeholders that the company will meet its commitments to environmental and social responsibility.

According to Neil Hawkins, corporate vice president, EH&S, and chief sustainability officer, resilience is an integral part of Dow's sustainable business practices and contributes directly to enterprise reliability. From a tactical perspective, businesses must strive to avoid disruptions due to turbulent change, including EHS incidents. In this case, resilience involves minimizing operational risks and being prepared to cope with a variety of unusual events, ranging from damaging floods to infectious disease outbreaks. Dow's global operations have survived many unexpected disruptions, including Hurricane Katrina in 2005 and the Japanese tsunami in 2011, thanks to well-developed business continuity practices.

On a more strategic level, Dow has been investigating how the company's operating sites can improve their resilience through a symbiotic relationship with the surrounding natural systems. The subject of ecosystem services has garnered increasing attention in the business community as scientists and economists have revealed the critical dependencies of our economy on commodities and functions that are furnished by nature. Ecosystems provide a continuous supply of freshwater, food, fiber, and other commodities that we tend to take for granted. Likewise, ecosystems provide critical functions such as flood control, climate regulation, and pollination, which ensure the stability of both markets and industrial supply chains. Only when disasters occur do we recognize how costly it might be to lose these services from nature.

Although many companies have begun to explore the value of ecosystem services, Dow took an unprecedented step in 2011, forming a collaboration with The Nature Conservancy (TNC). This initiative is aimed at helping Dow and the business community recognize, calculate, and incorporate the value of nature into business decisions, strategies, and goals. The Dow-TNC collaboration is based on the premise that consideration of ecosystem services in business decisions will improve both business performance and natural resource conservation.

The seeds of this bold endeavor were planted in 2008 when Dow's chairman and CEO, Andrew N. Liveris, challenged Hawkins to identify any elements missing in Dow's sustainability strategy. After much reflection, he identified ecosystem services as a relatively untouched issue: both a risk and an opportunity that could significantly affect shareholder value and community well-being. This subject was poorly understood in the business world, however, so Hawkins and his team began to search for motivating examples in Dow's own experience. Luckily, they discovered an overwhelming success story that had been achieved more than a decade earlier.

Back in 1995, environmental engineers were experiencing regulatory compliance challenges with the tertiary wastewater treatment pond at Union Carbide's Seadrift plant in Texas, which was acquired by Dow in 1999. The cost of replacing the system with a sequencing batch reactor (SBR) was estimated to be about $40 million. One creative engineer, Mike Uhl, proposed a nontraditional solution: designing a wetland to accomplish the same function. At first, Uhl's idea was brushed aside (buying a swamp?), but when he estimated that the initial cost of the wetland would be less than $1.5 million and that the operational costs would be significantly less than the SBR system, management began to take him seriously. After pilot testing proved that the approach was viable, this "green infrastructure" solution was accepted, and it has been operating successfully ever since.

A recent retrospective study estimated that the net present value savings achieved by the Seadrift wetland are more than $250 million (2012 dollars) over the lifetime of the system. Moreover, the life cycle environmental footprint (resource use and emissions) is considerably lower than the SBR would have been, and even the land use is comparable when one considers the entire supply chain of the SBR.[19] The constructed wetland also offers benefits to the community because it provides habitat for deer, bobcats, and birds and creates educational opportunities for local schools. Finally, the wetland system is resilient in several ways: it is decoupled from price changes or shortages in materials and energy, it does not degrade over time if properly managed, it cannot fail suddenly, and it is not vulnerable to human error because it requires virtually no supervision or maintenance. On the other hand, unlike conventional engineered systems, it requires specialized scientific knowledge and cannot be standardized and replicated easily.

Despite the compelling business case, the Seadrift approach has not yet been adopted at other Dow facilities. That may be due partly to the modesty of engineers like Uhl and partly to the hurdles of adopting and customizing an unfamiliar technology. One barrier is simply the larger land area required to establish the wetland (more than 100 acres), which is not available at many plants. Nevertheless, Hawkins recognized the Seadrift case as a compelling proof of his belief in the potential value of ecosystem services. He nominated Mike Uhl for Dow's prestigious Sustainability Innovator Award, which he won in 2009, effectively becoming a company hero.

Today, the Dow-TNC collaboration is working at several Dow sites to develop similar pilot projects. Much of the work is taking place at the sprawling industrial complex in Freeport, Texas, Dow's largest integrated manufacturing site and the largest single-company chemical complex in North America. It is located on the Gulf of Mexico amid a network of freshwater, marsh, and forest ecosystems that are critical not only to Dow's operations but also to fish and wildlife, agriculture, and local communities.

Initial results at Freeport have shown that Dow can harness ecosystem services to improve operational resilience in several ways, such as by improving air quality through reforestation, mitigating coastal hazards with natural infrastructure, and preventing disruption to freshwater supplies. For example, the team found that reforestation would be cost-competitive with conventional pollution control methods for ozone-forming nitrogen oxides. Dow has approached the State of Texas to allow reforestation as an ozone compliance measure in the state implementation plan. In addition, reforestation would sequester carbon dioxide, thus mitigating the adverse effects of climate change. If such forests are eligible for carbon offset credits, the associated costs could be substantially reduced (figure 7.2).

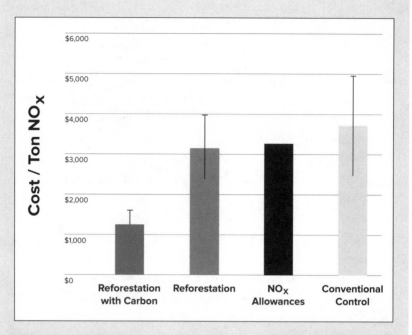

Figure 7.2. *Estimated removal cost per ton of nitrogen oxide (NO$_x$), comparing reforestation (with and without carbon credits) against NO$_x$ allowances and conventional pollution control alternatives*

Based on these encouraging results, Dow and TNC are developing a tool that can be used for purposes of financial decision making, known as the Ecosystem Services Identification and Inventory (ESII, pronounced "easy") Tool. The tool will allow companies to estimate the economic benefits that they derive from ecosystem services at a specific operating site as well as the value to surrounding communities from company-owned lands. The tool will initially cover eight ecosystem services: air quality control, climate regulation, erosion control, flood hazard mitigation, water provisioning, water quality control, water quantity control, and aesthetics.

One of the principal emerging threats to business continuity is shortages in freshwater supplies. Recognizing the inherent uncertainties in meeting its considerable water needs, the Freeport site managed to reduce its water usage by 10 percent per year beginning in 2012, thanks to the leadership of Earl Shipp, vice president of U.S. Gulf Coast operations. Dow is also working with TNC to develop modeling tools that can estimate the cost of water under various future climate scenarios, which may bring droughts in some locations and excessive erosion or flooding in other locations. Dow's intent is to develop standard work processes that incorporate these analytic tools into routine decisions, such as capital allocation, manufacturing technology improvement, and new product development.

Dow's ultimate vision is that the value of nature will be considered routinely as an integral part of business decision making. The company has been collaborating with other companies that share this vision, including Shell, Unilever, and Swiss Re. Together with TNC, these companies released a joint-industry white paper in 2013 entitled "The Case for Green Infrastructure."[20] They concluded that hybrid approaches, using a combination of green and gray infrastructure, may provide an optimum solution to a variety of shocks and improve the overall resilience (and reliability) of industrial business operations. In effect, natural systems provide a buffer that can absorb the inevitable fluctuations caused by climate change and socioeconomic turbulence.

Meanwhile, Dow is continuing to pursue product and process innovations that deliver improvements in both sustainability and resilience. Responding to the challenge of freshwater availability, Dow has introduced commercial innovations ranging from drought-resistant crops to more efficient water desalination technologies. Thus, understanding resilience is not just a defensive strategy for operational reliability; it also helps identify new market opportunities that will strengthen the resilience of Dow's customers and the communities they serve around the world.

Takeaway Points

- Traditional security practices that focus on physical and intellectual asset protection are now expanding to consider the security of critical environmental resources (e.g., soil, water, minerals).

- Environmental threats to enterprise resilience include natural disasters, deliberate attacks, regulatory and market pressures, resource scarcities, and loss of ecosystem services.

- Companies can shore up their resilience by improving environmental systems and processes, ranging from closed-loop resource recovery to product stewardship.

- There is an urgent need for enterprises to collaborate with national and regional governments to address environmental vulnerabilities and move toward global environmental security.

- **Resilience in Action:** Dow Chemical has demonstrated that the use of "green infrastructure" can be less costly, more-resilient, and more environmentally beneficial than traditional wastewater treatment practices.

CHAPTER EIGHT

Organizational Resilience

Most companies are far better at executing their current activities than at adapting to long-term changes in the business environment. Very few can do both well.

Eric Beinhocker[1]

The previous chapters have addressed enterprise resilience mainly from a functional and structural perspective, focusing on key business processes and interdependencies with customers, suppliers, and the broader environment. If we consider an enterprise to be a living system, however, it is important to consider its living components: human beings. The attitudes, skills, and behavioral norms of company employees are a vital determinant of the company's overall resilience. Although there has been a great deal of research on the resilience of individuals to stress and adversity, much less is known about the resilience of groups such as functional departments or informal teams. At a broader scale, there is growing interest in community resilience, which will certainly influence the resilience of companies that have operating facilities within a community. In times of crisis, companies, communities, and government agencies must all join forces to respond, recover, and forge ahead.

Companies that wish to improve their resilience will most likely need to invest in the structural aspects of enterprise resilience such as reserve capacity, flexibility, redundancy, and external alliances. They will also need to rethink their functional processes such as forecasting, trend analysis, risk management, and crisis management. These improvements, however, will not help in overcoming unexpected challenges unless the workforce is appropriately trained, equipped, and empowered to transcend business as usual and make the right decisions at critical times. Organizational resilience is the essential ingredient that enables companies to carry out effective resilience strategies: sensing change, adapting effectively, and coping with sudden disruptions.

There have been many decades of research in the field of organizational behavior on how to maximize human resource effectiveness, including management techniques, reward systems, and motivational strategies. In the fields of medicine and psychology, there is a vast amount of knowledge about resilience in individual behavior, including tenacity, self-confidence, and capacity to cope with adversity. More recently, in the field of sociology, there has been a wave of interest in the resilience of communities that are recovering from disasters. Although an in-depth review of all this knowledge is beyond the scope of this book, this chapter summarizes recent work that is specifically aimed at understanding the human dimensions of enterprise resilience.

Resilience, Strategy, and Culture

Businesses today invest a great deal of effort in strategy development and strategic planning. Any given strategy rests on several situational assumptions about markets, competition, and technology; therefore, the strategy is only effective during the window of time when these assumptions remain valid. Resilience is different: it is not a transient strategy but a permanently desirable attribute of the enterprise. The resilience strategies described in chapter 1 are universally applicable, regardless of industry type and business conditions. Of course, situational awareness is still important because it influences the tactical question of how to develop the right portfolio of resilience capabilities.

For an enterprise to survive and flourish, resilience needs to be embedded into the culture of the organization. Naturally, there are many cultural barriers to organizational resilience. During periods of successful performance, organizations may not be motivated to work on improving their resilience due to the inertia of the entrenched bureaucracy, complacency with the status quo, and overconfidence in the current enterprise strategy. These human tendencies allow organizations to drift into a false sense of security until a shock for which they are unprepared occurs.

Conversely, during times of stress and disruption, organizations may try to take short-term remedial actions and minimize the cost of recovery without considering what underlying vulnerabilities are the root causes of the disruption. Some organizations may adopt a fatalistic posture, believing that crises are inevitable and that the costs of prevention are too great. In today's business environment, with rising insurance costs and more stringent exclusions, such attitudes have become untenable.

It is common for companies to institute fundamental transformations in organizational structure and strategic emphasis, often corresponding to a transfer of senior leadership or corporate ownership. Such moves can help a company revive employee enthusiasm and adapt successfully to changes in markets and technologies, as repeatedly demonstrated by IBM and other long-lived companies. On the other hand, organizational change can be distracting and risky; an obvious example is BP's effort to look "beyond petroleum" and invest in renewable energy, which may have contributed to its neglect of basic operational safety. Also, employees may become weary of repeated change initiatives and may view "resilience" cynically as just the latest buzzword.

To avoid negativity, Liisa Välikangas suggests that companies should emphasize fitness and adaptability during periods of successful performance. According to Välikangas, that is the best time for building the capacity to deal with unexpected disruptions: "Anticipating certain challenges and preparing for them is one thing; the more challenging aspect of resilience is to be ready for emergent challenges coming our way . . . even beyond those changes that have been correctly anticipated."[2]

Gary Hamel and Välikangas describe the quest for resilience as seeking "zero trauma."[3] In today's world, that may be an unrealistic expectation, but it is certainly a worthwhile goal. To build resilience will require practice, experimentation, and rehearsal. It is a necessary step toward developing fitness so that in times of stress the organization can operate with skill and confidence. Just as military organizations engage in war games, resilient enterprises use exploratory exercises to pull employees away from the status quo and engage them in hypothetical problem solving and innovation.

The concept of mindfulness was identified by Karl Weick and Kathleen Sutcliffe as a typical characteristic of high-reliability organizations.[4]

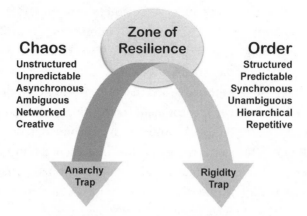

Figure 8.1. Finding the balance between order and chaos

Rather than following a mechanical procedure, mindful workers continually evaluate the situation, take note of small failures, avoid oversimplification, and remain sensitive to changing conditions. They anticipate undesirable events before they occur and take effective action to contain such events if they do occur so as to preserve or recover functionality.

A particularly striking view of organizational resilience based on the notion of structured chaos was developed by Shona Brown and Kathleen Eisenhardt.[5] They claim that companies such as Microsoft, 3M, Nike, and Intel have learned to embrace change by finding the right balance between structure and chaos, as depicted in figure 8.1. An organization that is overly structured and clings to stability may be unable to respond to change and can descend into a rigidity trap where it is no longer competitive. At the other extreme, an organization that is overly chaotic and lacks cohesion may descend into an anarchy trap where it is unable to survive.

Attributes of a Resilient Organization

In a resilient organization, the capacity to adapt to change and recover from crises becomes part of everyone's job. It is no longer just the responsibility of the professional groups that handle safety, security, risk management, and emergency response. It is embedded in the corporate DNA. Establishing a culture of resilience may not be easy, though, especially for large organizations. How can a company develop an organization that is resilient and competitive in the face of change? What are the important attributes that should be cultivated? These questions are particularly challenging for multinational companies expanding into developing nations, where the characteristics of the workforce are very different from those in domestic communities.

Organizational resilience is closely tied to organizational culture and strategic management of innovation. For example, Välikangas defines the following five dimensions of organizational resilience:[6]

1. **Organizational intelligence:** Accommodating multiple voices and diversity of thought

2. **Resourcefulness:** The ability to innovate to mitigate risk or overcome constraints

3. **Robustness:** Structural design to support resilient behavior and avoid systemic traps

4. **Adaptability:** Fitness for change based on experimentation and rehearsal (see above)

5. **Tenacity:** Determination to rise to a challenge rather than accepting adversity or defeat

Similarly, Patricia Longstaff identifies key attributes of resilient organizations that operate in an uncertain environment.[7] Given the complexity and unpredictability of large organizations, improvisation and deductive tinkering are necessary to cope with surprises and lack of information. In times of crisis, people may need to depart from the standard operating procedures, but, unfortunately, such behaviors are often punished rather than rewarded. Companies need to establish a culture of flexibility, trust, and openness so that employees are not deterred from innovative thinking.

The attributes that determine organizational resilience are difficult to measure because they are more abstract than functional or structural characteristics. The only practical way to test disaster resilience is through simulated disruptions because actual events are, one hopes, rare. It is even more difficult to test the positive aspects of resilience in terms of seizing emergent opportunities. Instead, many practitioners have devised assessment schemes that measure the attributes of a resilient organization using qualitative instruments, similar to the SCRAM approach described in chapter 6.

SCRAM includes the following basic elements of organizational resilience within various capability factors:

- The organization emphasizes the **empowerment** of on-site experts to make key decisions, regardless of level of authority.

- The organization has the ability to **collaborate** effectively with others for mutual benefit, both internally and externally.

- The organization encourages creative **problem solving**, thus enabling resourcefulness in times of stress or crisis.

- The organization enforces individual **accountability** for performance, thus encouraging results-oriented employee engagement.

- The organization encourages **diversity** in personal backgrounds, skills, and experience, thus increasing versatility and robustness.

- The organization is capable of filling **leadership** voids quickly, implying attention to career development and succession planning.

- The company strives to be a **learning** organization, regularly using feedback and benchmarking tools to learn from past experience and from other organizations.

- The organization has a culture of **caring** for employees, thus building loyalty, solidarity, and teamwork that will serve the company well in times of stress or crisis.

- The organization has strong, long-term **relationships** with each of its business-to-business customers and communicates effectively with all customers.

In a parallel effort, a group of university researchers in New Zealand has developed a framework for assessment and management of organizational resilience.[8] The framework consists of thirteen qualitative resilience indicators grouped into three major categories: leadership and culture, networks, and change readiness.

Leadership and culture

- **Leadership:** Strong crisis leadership provides good management and decision making during times of crisis as well as continuous evaluation of strategies and work programs against organizational goals.

- **Staff engagement:** The organization engages and involves staff to understand the linkages between their own work, the organization's resilience, and its long-term success so that they are empowered and use their skills to solve problems.

- **Situation awareness:** Staff are encouraged to be vigilant about the organization, its performance, and potential problems. Staff are rewarded for sharing good and bad news about the organization, including early warning signals, and such information is quickly reported to organizational leaders.

- **Decision making:** Staff have the appropriate authority to make decisions related to their work, and authority is clearly delegated to enable a crisis response. Highly skilled staff are involved, or are able to make, decisions when their specific knowledge adds significant value or when their involvement will aid implementation.

- **Innovation and creativity:** Staff are encouraged and rewarded for using their knowledge in novel ways to solve new and existing problems and for using innovative and creative approaches to developing solutions.

Networks

- **Effective partnerships:** The organization develops an understanding of the relationships and resources that it might need to access from partner organizations during a crisis and carries out planning and management to ensure this access.

- **Leveraging knowledge:** Critical information is stored in several formats and locations, and staff have access to expert opinions when needed. Roles are shared, and staff are trained so that others will always be able to fill key roles.

- **Breaking silos:** The organization minimizes divisive social, cultural, and behavioral barriers, which are most often manifested as communication barriers that create disjointed, disconnected, and detrimental ways of working.

- **Internal resources:** The organization manages and mobilizes its internal resources to ensure its ability to operate during business as usual as well as being able to provide the extra capacity required during a crisis.

Change readiness

- **Unity of purpose:** The organization ensures broad awareness of what the priorities would be following a crisis, clearly defined at the organization level, as well as an understanding of its minimum operating requirements.

- **Proactive posture:** The organization establishes strategic and behavioral readiness to respond to early warning signals of change in the internal and external environment before they escalate into crisis.

- **Planning and strategies:** The organization develops and evaluates plans and strategies to manage vulnerabilities in relation to the business environment and its stakeholders.

- **Stress-testing plans:** Staff participate in simulations or scenarios designed to practice response arrangements and validate plans.

Table 8.1 summarizes the various organizational dimensions, attributes, indicators, and factors identified above and shows how they

Table 8.1. Organizational resilience factors corresponding to fundamental enterprise attributes

Organizational Resilience Factor	Enterprise Attribute
Agility, proactive posture	Adaptability
Innovation and creativity	
Internal resources	
Learning, feedback	
Resourcefulness, problem solving	
Situation awareness	
Breaking silos	Cohesion
Collaboration, partnerships	
Leadership, staff engagement	
Tenacity, toughness	
Trust, openness, caring	
Unity of purpose	
Decision making	Efficiency
Leveraging knowledge	
Organizational intelligence	Diversity
Robustness, versatility	
Staff empowerment, accountability	

correspond to the fundamental attributes of resilience defined in chapter 5: adaptability, cohesion, efficiency, and diversity.

Improving Organizational Resilience

How can organizations cultivate the attributes necessary to improve their resilience? The first step is to look internally at the corporate culture and behavioral norms. A process similar to that described in chapter 6 for supply chain management can be applied to human resource management. In fact, the process is even simpler because these organizational attributes are not closely tied to specific vulnerabilities. A straightforward capability assessment will reveal gaps or weaknesses, and then the organization can select high-priority areas for improvement. Broadly speaking, there are several principles that companies can follow to foster resilience in the workforce:[9]

- In employee recruitment and career development, emphasize creative skills, questioning of assumptions, decisiveness under uncertainty, and intellectual diversity.

- In organizational structure and processes, emphasize open communication and cross-functional collaboration, flexible job descriptions, continuous learning, and tolerance for experimentation.

- In cultural context setting, emphasize broad internal and external interactions, a climate of trust and interdependence, and respect for value contribution rather than hierarchical rank.

Moving beyond the familiar realm of human resource management, companies need to look outward to leverage their relationships with other firms and with society at large. Many companies have already recognized the power of crowdsourcing and social media as tools for expanding their knowledge and influence. Because of the interdependence of companies with their suppliers, customers, communities, and infrastructure systems, they cannot operate as isolated entities. As suggested in chapter 4, an enterprise is part of a larger "system of systems," and its resilience is intertwined with the resilience of these systems.

The 3V framework introduced in chapter 3 helps clarify the interdependence between an enterprise and its many stakeholder groups. Figure 8.2 shows how the creation of shareholder value depends on the availability of talent—that is, human capital—as well as social capital:

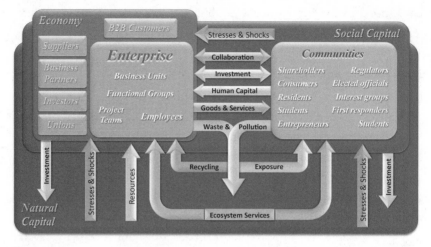

Figure 8.2. *Systems view of organizational resilience: Resource flows and interdependencies*

the institutions, relationships, and norms that underpin human society, including bonds of mutual trust. Likewise, community well-being depends on the availability of commercial high-quality goods and services as well as high-quality jobs. Both enterprises and communities may be disrupted by stresses and shocks, including natural disasters, technological failures, and social or political upheavals.

Organizational resilience involves not only capacity building and teamwork within the enterprise, but also external communication and collaboration with economic actors and community stakeholders. To ensure their resilience, enterprises and communities must invest in one other as well as in waste mitigation and protection of natural capital (see chapter 7). Particularly when it comes to anticipating and coping with disruptions, it is important for companies to join forces with a network of collaborators based on both formal and informal organizational connections.

Graceful Extensibility

A research group at Ohio State University led by David Woods has been studying the phenomenon of resilience in a variety of different organizations. The group's approach, called resilience engineering, identifies fundamental principles and practices that enable organizations to cope with complexity, adapt to change, respond effectively to unexpected disruptions, and sustain their required performance.[10] The characteristics of resilient organizations include the following:

- Experience in coping with surprising events, including major and minor disruptions
- Heightened awareness of potential vulnerabilities despite a record of past success
- Empowerment of the workforce to take initiative in response to unforeseen demands
- A sense of reciprocity that promotes mutual support across organizational boundaries

These characteristics all contribute to what Woods calls "graceful extensibility," the opposite of brittleness, providing organizations with the capacity to stretch beyond their normal operating boundaries to handle emerging challenges or crises.

Woods cites the example of an aviation transportation firm that has developed graceful extensibility due to the day-to-day pressures of its business, including airport congestion, weather delays, and fluctuating demand. The company adapts continuously to changing conditions by adjusting its schedules and fleet utilization. It must cope with peak demands during critical holiday periods as well as unusual stresses during extreme weather events such as hurricanes and snowstorms. As a result, it has evolved a horizontal operating style that avoids hierarchical decision making, thus encouraging functional managers to coordinate effectively and act in a timely manner and emphasizing responsiveness to customers over minimization of costs. Through daily practice, management has developed the skills to cope with turbulent change, so they are well prepared for decisive action if and when major challenges should arise.

In contrast, organizations that operate in a more stable business environment may grow overconfident, having never experienced severe stresses or cascading failures. Their inability to adapt effectively is only revealed when a surprise occurs, such as a computer system failure, and they may discover that their existing recovery plans and resources are inadequate. Having never really tested their capabilities, they are unable to keep pace with a rapidly escalating crisis, and their performance may completely unravel. To avoid this type of breakdown, Woods advises companies to build adaptive capacity and practice repeatedly in real or simulated situations. By investing wisely in improving their resilience, they will be better prepared not only to cope with downside risks, but also to seize new business opportunities that may suddenly arise

Resilience in Action

Emergency Response at American Electric Power[11]

When the electricity doesn't work, it is not just the lights that go out. Information, communications, transportation, and water and sewer networks all depend on the availability of electric power at some point in their production or delivery process. Virtually all service providers and every retail point of sale in the country depend on electricity.

As a result, the $220 billion electric power industry has become best in class at recovering from the localized, usually weather-related, disruptions that affect every region of the United States. Every year, electric utilities must handle major power outages due to ice storms, hurricanes, fires, tornadoes, and blackouts. Although the situation and scale of an outage is different each time, the emergency preparedness and response procedures are very similar.

Other industry sectors can learn from the electric power industry how to prevent, prepare for, and respond quickly to threats and supply disruptions, including deliberate attacks. Thus, electric utilities provide a useful model for enterprise resilience. A recognized leader in the field of emergency response is American Electric Power (AEP), one of the largest electric utilities in the United States, delivering electricity to more than five million customers in eleven states. AEP owns more than 36,000 megawatts of generating capacity as well as the nation's largest electricity transmission system, spanning nearly 39,000 miles.

Responding to Severe Weather Events

AEP's resilience was tested on January 12, 2006, when a severe ice storm struck several communities in the territory served by Public Service of Oklahoma (PSO), an AEP operating company based in Tulsa. The storm came in three successive waves over a period of several days, depositing up to 2 inches of ice. PSO's crews were forced to delay line repairs because they knew that the lines would most likely break again in the next wave of ice. Ultimately, the storm interrupted electrical service for close to 250,000 customers, about 60 percent of the area's population. The hardest-hit community was McAlester, which lost power for more than ten days.

To respond to such disruptions, AEP has evolved an elaborate, company-wide system governed by a detailed service restoration plan that is updated continually. Corporate oversight is provided by AEP's manager of emergency restoration planning, who at that time was Jim Nowak. He convened an ice-event conference call twenty-four hours before the Oklahoma storm hit. During the Oklahoma event, PSO requested assistance from other AEP operating companies as well as neighboring utilities and was able to mobilize more than two thousand emergency workers promptly.

The coordination required to manage and support these emergency resources is an enormously complex task. Outside contractors are often used; AEP contracts with forestry companies to clear branches for the line crews and with base logistics companies to supply tents, trailers, food, and laundry services. AEP has adopted advanced technologies, such as handheld data entry and communication devices, to help dispatch crews quickly to the areas of greatest need. Satellite positioning devices are installed on line repair trucks so that resources can be monitored centrally and deployed in real time.

The service restoration plan sets out a detailed organizational structure, with different levels of responsibility. Voluntary participation—all hands on deck—is part of the AEP culture, and during an emergency it is not unusual for more than 75 percent of employees in the affected operating company to be engaged. Each person receives an alternative "storm" assignment; for example, hazard standby associates are assigned to guard broken wires to prevent residents from being injured. AEP provides standardized training and materials so that different operating companies can collaborate effectively.

PSO's extraordinary performance during the 2006 ice storm was acknowledged by local officials. The superlative emergency response capabilities of AEP were recognized by Edison Electric Institute and by other industry and government organizations, including the Department of Defense.

Mutual Assistance for Emergency Response

When a major event produces widespread outages, the electric industry mobilizes to deliver resources, supplies, and crews needed to get the lights back on safely and quickly. This practice of mutual assistance, which dates to the 1950s, helps utilities mitigate the risks and costs of major outages through sharing of resources. The utilities that seek assistance pay the costs of the utilities and contractors providing labor and equipment. It is common for AEP and other utilities to provide emergency support to one another, as was the case in the ice storm that affected PSO.

For example, when four major hurricanes hit the southern states in 2004, AEP quickly moved qualified employees to the affected areas. Because of the scale of the recovery efforts and the mixture of different safety cultures and work standards, the company had to change its mutual assistance procedures. Even while the Hurricane Charlie response was taking place, AEP's emergency restoration planning group was forming a new framework to prepare for other hurricanes. With Hurricane Frances looming, the company had a plan formulated and approved eight days after Charlie was over. Included in the new plan was a percentage-based formula to determine the number of employees needed to respond to restoration work while providing enough staff to service the needs of the company at home. The process also included a daily conference call that included a status report on each crew working on potential problems and plans for the next day.

AEP's emergency response experience carries over into business continuity planning and stimulates consideration of different disruption scenarios, such as unavailability of facilities, systems, and employees. The sources of a service disruption could vary from unplanned events such as accidents or storms to deliberate threats such as theft or vandalism. AEP engages in detailed planning processes and drills to prepare for such threats within the company as well as in broader power pools (such as the Pennsylvania, New Jersey, Maryland power pool and the Southwest power pool). The overall imperative is to keep capacity broadly available among the partners. One of the greatest concerns is "cascading failures," such as the Northeast blackout of 2003 that affected 55 million people in the United States and Canada.

To improve coordinated response to power interruptions affecting multiple US regions, AEP worked with more than thirty other utilities to form the National Response Event (NRE) framework in 2013. The NRE has created a national mutual assistance resource team that will pool and allocate resources from regional mutual assistance groups. When a major event occurs, these groups will act in unison to ensure the highest level of resource coordination.

Meanwhile, based on the lessons of the previous decade, AEP developed a new emergency response plan (ERP) for implementation in 2015. This ERP was motivated by the critical reviews of utilities' restoration activities by regulatory commissions in New York, Maryland, and Connecticut after major Northeast storms. A key element of the ERP is establishment of an incident command system, a crisis management tool used by the Federal Emergency Management Agency and increasingly adopted by industry. This system helps improve management efficiency; reduce redundancy; define employee responsibilities; and improve communications with first responders, agencies, and customers.

Resilience Capabilities at AEP

Based on the SCRAM assessment methodology (see chapter 6), the following capabilities represent key strengths for American Electric Power.

Anticipation: To facilitate rapid response and recovery of service, the mobilization process usually starts in advance of a major disruption. Availability of resources, deployment schedule, and risk mitigation strategies are identified early in the process.

Adaptability: Depending on the scale of the disruption and the potential of future impacts, service restoration procedures are often modified in real time.

Capacity: AEP takes advantage of the system-wide resources from its operating companies, coordinated through the corporate office. In addition, excess "surge" capacity is made available through mutual assistance networks as described above.

Collaboration: Operating procedures and equipment standards are consistent among all AEP operating units, as is the training of all employees. This common foundation allows for effective real-time communication between units during unanticipated disruptions.

Dispersion: AEP benefits from the multiplicity of resources that are dispersed across the seven different operating companies. The service territories tend to spread over broad areas rather than being concentrated in large cities.

Recovery: A framework for effective communication, including conference call formats for sharing weather forecasts, damage estimates, and labor requirements, has been established to coordinate information exchange during emergency response and restoration efforts.

Visibility: As mentioned above, AEP has adopted technologies such as handheld data entry and communication devices to expedite the dispatch of repair crews to the areas of greatest need. Satellite positioning devices on line repair trucks enable resources to be centrally monitored and deployed in real time.

Moving toward Grid Resilience

Following the 2003 Northeast blackout, the North American Electric Reliability Corporation (NERC) was authorized by the Federal Energy Regulatory Commission to enact and enforce rules and standards that would advance the reliability and resilience of the bulk electricity system. In 2013, NERC partnered with the utility industry to form the Reliability Assurance Initiative, a collaborative process that identifies reliability risks and uses that information to better gauge future compliance monitoring and enforcement efforts. The aim of regulators is to have electric power companies monitor their own activities, detect issues when they occur, assess the risk of those issues, and correct the causes of those issues in a timely manner. In partnership with ReliabilityFirst, AEP has participated in pilot exercises to help refine NERC's monitoring, audit scoping, and enforcement processes.

This new approach signals a shift from a passive compliance culture to a more proactive resilience culture. To improve the resilience of the grid infrastructure to severe weather events, AEP created a distribution storm-hardening strategy team, which recommended design changes to improve the ability of utility poles to withstand strong winds and ice accumulation. These hardening measures are predicted to increase the strength of electric structures by at least 25 percent with a nominal increase in cost. Other examples of grid resilience initiatives include decreasing the distance between poles, installing fault isolation devices, and deploying smart grid technologies. Several states have enacted legislation that enables utilities to recover the costs of reliability and resilience improvements.

AEP has developed an assessment tool to help determine where to deploy capital funds to maximize the benefits of grid-hardening initiatives. Among the criteria to be used are the number of customers served, the type of customer (how many on a particular circuit are considered "critical" customers, such as hospitals and nursing homes, law enforcement agencies, and water or wastewater facilities), the age of the poles, and the average duration of outages. AEP is also participating in the Electric Power Research Institute's Grid Resiliency Project, which will provide new tools and strategies to improve the distribution system's ability to withstand severe weather events.

According to Laura Thomas, former CRO of AEP, the company's emphasis on reliable service delivery is essential to assuring customer satisfaction because "AEP is part of every company's business continuity plan." Jim Nowak added, "Restoring power is not just a responsibility, it's a moral imperative."

Takeaway Points

- In addition to structural investments and functional processes, enterprises need to consider the human element, which is essential for sensing change, adapting effectively, and coping with sudden disruptions.

- Organizational resilience transcends strategy and is closely tied to the culture of the enterprise; it should be cultivated in periods of success rather than in response to crisis.

- Operating in the zone of balanced resilience requires a balance between structure and chaos, avoiding excessive rigidity on the one hand and excessive anarchy on the other hand.

- There are a variety of conceptual schemes for assessing the attributes of a resilient organization, all corresponding to the fundamental attributes of adaptability, cohesion, efficiency, and diversity.

- Mechanisms for cultivating organizational resilience include improvements in recruitment practices, internal processes and cultural norms, and interactions with external stakeholders.

- **Resilience in Action:** American Electric Power has a comprehensive plan for dispatching its own employees and collaborating with other industry and government organizations to respond to winter storm emergencies.

CHAPTER NINE

Tools for Managing Resilience

The future ain't what it used to be.

Yogi Berra[1]

The concept of resilience may seem intuitively clear, but it is daunting from an engineering, design, and management point of view because of its breadth and multifaceted nature. To incorporate resilience thinking into the day-to-day work of a business will require several supporting tools. The five main requirements for a resilience toolkit are as follows:

1. **Business innovation.** Teams working on strategic planning, product development, business process improvement, and other creative tasks need to understand basic resilience principles and need guidance on how to strengthen the resilience of their plans and designs.

2. **Screening and prioritization.** When comparing and contrasting different courses of action, business teams need methods and tools for judging the resilience of a proposed product design, process improvement, or system intervention relative to the status quo and competing plans or designs.

3. **Performance estimation.** When conducting in-depth evaluation of the strengths and weaknesses of alternative plans and designs, business teams need methods and tools to evaluate their anticipated performance, including resilience and other indicators.

4. **Decision making.** Managers need methods and tools for considering resilience alongside many other business objectives, weighing the

trade-offs in terms of costs, risks, and benefits for the enterprise and its stakeholders, and deciding on the best alternative.

5. **Performance evaluation.** Managers need methods and tools for tracking the results of their decisions, including analysis of whether their investments in resilience improvement yielded a positive return in terms of financial or other performance indicators.

Table 9.1 gives an overview of different types of tools and methods available for these purposes. Because the practice of enterprise resilience is still relatively new, many different approaches are emerging or are under development. This chapter provides a snapshot of current tools and methods that are relevant to these five major categories of needs.

Note that *resilience cannot be studied as an isolated phenomenon.* As shown in chapter 3, the resilience of an enterprise is dependent on other systems to which it is coupled. In a highly connected, turbulent world, adaptive management becomes more important than forecasting, and an enterprise must strive for resilience in both its internal processes and external linkages. Thus, an enterprise needs to understand the resilience of the people it employs, the communities in which it operates, the infrastructure systems on which it depends, the suppliers and partners with which it collaborates, and the markets that it serves. As discussed in chapter 11, many of the methods and tools described below are applicable not only to enterprises, but also to their external partners.

Table 9.1. Potential components of a resilience toolkit

Activities	Qualitative Tools	Quantitative Tools
Business innovation	Guideline checklists Examples of resilience	
Screening and prioritization	Criteria checklists Scoring methods	Leading indicators Vulnerability analysis
Performance estimation	Qualitative indicators	Performance indicators Predictive simulation Stress testing
Decision making	Scoring methods Scenario-based planning	Risk analysis Cost-benefit analysis
Performance tracking		Lagging indicators Return-on-investment analysis

Qualitative Tools

Qualitative tools and methods have obvious advantages over quantitative methods: they are easier to apply, require minimal data, and can be useful despite large uncertainties. In the early stages of planning and innovation, qualitative methods are preferable because they can provide valuable insights without requiring large resource expenditures. The following are general types of qualitative tools that can be adapted to the needs of individual companies.

Guidelines are helpful for codifying knowledge about resilient design strategies. They can range from general principles, such as "ensure that backup electric power is available," to very specific instructions about how to perform a resilience audit for a manufacturing facility. Chapter 10 discusses the process of designing for resilience and gives several examples to illustrate how such guidelines can be applied.

Checklists are the simplest and most widely used form of qualitative tool. For example, decision criteria are often disseminated as a checklist of points to consider, sometimes stated in the form of questions (e.g., "Were alternative forms of transportation considered?"). Checklists have the advantages in that they require only modest resources to update and maintain and that they are easy to understand and implement. Chapter 11 offers an example of an energy resilience checklist for communities.

Despite their advantages, checklists have important limitations:

- Checklists are both crude and subjective and do not capture the range of possible situations. A simple answer of "yes" or "no" conveys little information and cannot be assigned much confidence. Therefore, they can provide only vague indicators of resilience characteristics. For example, a supplier checklist might pose the question "Do you have a waste minimization program?" A more detailed assessment would take into account the baseline waste stream, the types of wastes, the difficulty of recycling, and the level of improvement achieved.

- Multiple checklists that reflect different perspectives can produce conflicts. For example, a resilience checklist may call for maintaining a reserve of critical supplies, whereas a materials management checklist may call for eliminating unnecessary inventory.

- Checklists provide no guidance about the relative importance of different resilience issues or the degree of effort that is warranted in addressing a specific issue. For example, is it more important to provide redundancy or to design for modularity? Such challenging questions can only be answered through a more rigorous trade-off analysis.

- Checklists can actually reduce creativity by creating a false sense of security. Having worked through a resilience checklist in mechanical fashion—literally checking the boxes—employees may take comfort in the belief that they have adequately considered resilience issues. They may not become truly engaged and may overlook important opportunities or problems that are not on the checklist.

Checklists are nevertheless an effective starting point for encouraging organizations to think about resilience issues and to begin taking positive actions.

Scoring methods are another type of qualitative tool, more sophisticated than simple checklists. They can be used for purposes of screening and comparison and can even be helpful in trade-off analysis for decision making. The SCRAM resilience index described in chapter 6 is an example of a scoring method that can be applied to a business unit or product category to identify strengths and weaknesses. Such methods are particularly useful for qualitative attributes, such as social capital, or in cases in which the lack of adequate data and models makes quantitative assessment impossible.

Scoring methods are often used to compare several alternative designs or plans. They involve creating a matrix diagram in which the rows represent different alternatives and the columns represent attributes of interest. A scoring protocol can then be applied based on available data or subjective judgments to derive categorical or numerical ratings. The assigned scores are seldom physically meaningful in an absolute sense, but they can be used to distinguish the *relative* strengths of available alternatives. There are many variations on this basic technique, including scorecards, rating schemes, and "traffic light" signal charts.

Although scoring methods are popular, some caveats are in order. A mechanistic, repeatable scoring algorithm will lead to conclusions that are at best approximate and occasionally just wrong. Another weakness of such methods is that they tend to aggregate the results into a single meaningless number. Without delving into the logic of the system, it is difficult for planners and designers to understand what changes might result in an improved score.

Scenario-based planning methods are helpful for identifying enterprise vulnerabilities, anticipating potential disruptions, and developing strategies for resilience. The complacent approach to strategic planning and business forecasting is simply to use the previous year as a baseline and then estimate how performance may change due to current trends. In most industries, this approach is a recipe for disappointment. When considering broad systemic changes, such as macroeconomic trends, precise forecasting is usually not possible.

Instead, recognizing the extreme uncertainties in the business environment, many companies have adopted planning methods that question old assumptions, envision alternative future scenarios, and then use "backcasting" to develop strategies for success under different possible circumstances. For example, Royal Dutch Shell has been a pioneer in using scenarios to support long-term strategic planning (see "Shell's Future Scenarios" box). Another example, involving several federal agencies, was the application of a "foresight" approach, which envisioned a 2040 scenario of energy security in the United States and used backcasting to formulate a grand strategy for energy security that could be adopted today.[2]

Shell's Future Scenarios[3]

The latest iteration of Shell's future scenarios proposes two possible lenses for envisioning the world of tomorrow, looking ahead fifty years or more. The first is the **mountains lens**, which envisions a world in which *status quo* power is locked in and held tightly by the influential elites. Here, stability is the highest prize: those at the top align their interests to unlock resources steadily and cautiously, not solely dictated by immediate market forces. The resulting rigidity within the system dampens economic dynamism and stifles social mobility.

The second scenario is the **oceans lens**, which envisions a world in which influence stretches far and wide, power is devolved, competing interests are accommodated, and compromise rules. Economic productivity surges on a huge wave of reforms, yet social cohesion is sometimes eroded and politics destabilized. Much secondary policy development then stagnates, giving immediate market forces greater prominence.

These alternative scenarios highlight the continuing tensions between flexibility and control, wealth and equality, order and chaos. For Shell, the scenarios provide insights into the drivers of energy supply, demand, regulation, and innovation and help prepare the company for many possible pathways of global economic development. This approach exemplifies systems thinking in support of enterprise strategy.

Resilience Indicators

Managing enterprise resilience requires the use of qualitative or quantitative indicators, also known as metrics. The old adage that "you can't manage what you can't measure" certainly holds true for resilience. As shown in table 9.1, there are generally three possible uses for indicators:

1. Characterizing the *attributes* of alternative plans or designs before they are implemented

2. Estimating the *predicted performance* of plans or designs before they are implemented

3. Assessing the *actual performance* of plans or designs after they have been implemented

The underlying assumption here is that the performance of the business will be raised by improving selected resilience attributes (e.g., adaptability). Performance can be measured in many ways, including monetary returns (e.g., net profits), operational results (e.g., percent on-time delivery), and intangible outcomes (e.g., customer retention). Most businesses and functional departments identify a small set of key performance indicators (KPIs) that are used for decision making and reporting purposes. The concept of a balanced scorecard is a popular approach for selecting KPIs.

The following list of *selection criteria* can be used to choose key performance indicators.[4] The set of indicators should be

- *Relevant* to the interests of the intended audiences, including advancement of business interests for shareholders and management and enhancement of social and environmental outcomes for concerned stakeholders

- *Meaningful* to the intended audiences in terms of clarity of indicator definition, comprehensibility, and transparency

- *Objective* in terms of measurement techniques and verifiability while allowing for regional, cultural, and socioeconomic differences

- *Effective* for supporting benchmarking and monitoring over time as well as making decisions about how to improve performance

- *Comprehensive* in providing an overall evaluation of the company's products and services and recognizing issues that can influence supplier or customer performance

- *Consistent* across different sites or facilities, using appropriate normalization and other methods to account for the inherent diversity of businesses

- *Practical* in allowing cost-effective, nonburdensome implementation and building on existing data collection where possible

Based on generally accepted accounting principles, an overarching criterion in selection of performance indicators is "materiality," the significance of performance results for purposes of decision making by management, shareholders, and other stakeholders.

Does resilience actually improve performance? Researchers at Ohio State University worked with a sample of eight companies to investigate the effect of supply chain resilience on performance. In this study, each participating firm provided up to twelve supply chain performance indicators, including availability, delivery lead time, inventory position, order accuracy, and customer complaints. (The supply chain resilience indicator used was the SCRAM qualitative index described in chapter 6.) Figure 9.1 demonstrates the study findings; in general, the higher the resilience

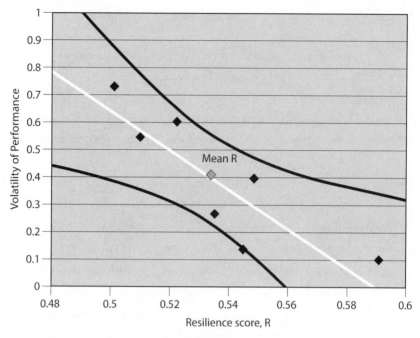

Source: Pettit, Croxton, Fiksel, *Journal of Business Logistics*, 2013

Figure 9.1. *Supply chain resilience reduces volatility of performance*

score, the lower the volatility of supply chain performance indicators.[5] This encouraging result suggests that resilience leads to more stable performance.

Resilience indicators can play two roles: leading or lagging. Leading indicators provide a predictive tool to anticipate changes in performance, whereas lagging indicators provide a retrospective view of actual performance. During the early stages, before plans or designs are firmly established, it is helpful to identify what resilience attributes the company wishes to emphasize as leading indicators. In the above example, we can infer that the SCRAM resilience score is a leading indicator of supply chain volatility.

Table 9.2 lists some typical attributes for which resilience indicators can be defined. Some of them can only be measured in qualitative terms, whereas others can be quantified. For example, recoverability can be measured in terms of the time required to recover, the cost of recovery, or the maximum tolerable degree of disruption. Also, these indicators may

Table 9.2. Examples of enterprise resilience attributes and possible indicators

Attribute	Types of Indicators	Example
Cohesion	Strength of corporate identity or stakeholder trust	Interbrand ranking of brand value
Vulnerability	Presence of disruptive forces that can threaten business continuity	Country-specific political risk index
Adaptability	Capacity to rapidly modify key products, technologies, or business processes	Response time to execute modification
Efficiency	Productivity in terms of value delivered relative to resources required	Production volume per unit of energy input
Diversity	Variety of markets, suppliers, facilities, and employee capabilities	Number of qualified sources by component
Stability	Ability to continue normal business operations when disruptions occur	Surge capacity as a percent of normal output
Recoverability	Ability to overcome severe disruptions and restore business operations	Maximum tolerable damage without shutdown

be correlated; for example, stability, vulnerability, and recoverability are all dependent on a fundamental attribute called precariousness,[6] which indicates how close the system is to a critical threshold (e.g., minimum inventory level). As noted in table 9.2, resilience indicators can also be used as lagging indicators to assess the actual resilience of the business.

For purposes of performance monitoring, it is often helpful to aggregate resilience indicators into an index, similar to SCRAM. For example, Cisco created a resiliency index that is a composite of resilience indicators related to products, suppliers, manufacturing processes, and test equipment for outsourced components. This index is applied automatically to Cisco's top one hundred products, accounting for about half of Cisco's revenue, and is included on the company's Supply Chain Operations Executive Dashboard.

Quantitative Analysis Methods

Quantitative methods are needed to enable systematic assessment and quantification of performance for a product, process, or asset. Assuming that indicators have been defined and objectives have been set, analysis methods can be used to support design decisions in several ways:

- To evaluate the degree to which a particular initiative meets cost or performance objectives and requirements
- To compare alternative approaches and evaluate their relative merits
- To identify potential improvements and evaluate their expected benefits

Quantitative methods are preferable for purposes of continuous improvement and competitive benchmarking. In many cases, quantification is difficult, however, and qualitative methods can serve adequately.

Resilience is a relatively new managerial concept, and the practice of resilience measurement is still at an early stage. Because resilience is holistic in nature, it cannot be easily disaggregated. The total cost or total energy demand of a complex system can be calculated by adding up the individual cost or energy demand of the system components. Resilience does not lend itself to this type of calculation because the resilience of the whole is dependent on the system-level design and the interactions among of the parts. For example, a supply chain is only as resilient as its weakest link, so quantifying supply chain resilience will require analyzing all stages of the product life cycle, including raw material extraction, processing, transport, component manufacturing, product assembly, distribution, end use, service, and end-of-life disposition, including recycling.

There are several major categories of quantitative analysis methods: predictive simulation, risk analysis, and cost-benefit analysis. Each is described in the following sections.

Predictive Simulation

To develop a business case for resilience, companies will need to consider the trade-offs between resilience factors and other important performance goals, such as cost, quality, manufacturability, and reliability. Resilience is closely linked with these other metrics. Sometimes there are synergies, as when reducing downtime lowers operating costs, and sometimes there are compromises, as when increasing flexibility requires capital investment. To understand these complex trade-offs fully may require the development of predictive models that estimate the effects of a proposed change.

Investments in improving enterprise resilience can be difficult to justify when the goal of these investments is to prevent hypothetical disruptions. In some cases, the likelihoods and magnitudes of such disruptions are quantifiable, so decision makers can weigh the costs against the benefits using the risk analysis methods described below. More often than not, there are large uncertainties about the timing and severity of potential disruptions, and it may be difficult to enumerate all the possible causes. For example, a work stoppage might be triggered by operator error, process failure, raw material shortage, weather disturbance, or union action. In such circumstances, it is still possible to build a business case for resilience by simulating a disruption and investigating how resilience strategies (e.g., maintaining multiple manufacturing locations) could mitigate the adverse business consequences. Generally speaking, predictive modeling and simulation tools can be used to assess the expected resilience of alternative enterprise configurations under varying assumptions and future scenarios.

Predictive simulation may involve considerable effort in developing a model of enterprise operations. The purpose of building the model is to enable estimation of changes in key performance metrics (e.g., order fulfillment rate) due to potential disruptions. Once developed, such a model can be used repeatedly to evaluate how different design options might improve operational resilience and performance. Some companies have developed internal modeling expertise and established modular simulation architectures that enable models to be customized rapidly to meet the needs of specific business units. The basic requirement of a modular architecture is that it uses standard components to produce valid and reliable simulation models. A library of reusable modules including various enterprise assets and disruption scenarios can be compiled to facilitate model building. Chapter 3 describes the development of predictive simulations based on the 3V model.

For example, development of a business-specific supply chain simulation would begin by characterizing the existing supply chain structure and identifying priorities for resilience improvement, perhaps based on a SCRAM analysis. The simulation model structure will then correspond to selected supply chain operations and resources of the business unit. Typically, several major classes of modules will be used to represent different classes of supply chain participants: suppliers, manufacturing sites, storage warehouses, transportation modes, and customers. Each class of modules will have standard operating characteristics, including time schedules, capacities, lead times, and inventory levels. Resilience modeling can then be performed by stress testing: simulating supply chain responses to business fluctuations or external disruption scenarios, such as a loss of capacity, a demand surge, or a transportation breakdown. This process is similar to the stress testing required of financial institutions to determine whether they can weather an economic crisis.

Dow Chemical frequently applies this type of simulation approach to support business divisions. For example, Dow developed a system dynamics simulation to help design the largest single chemicals complex ever to be constructed in one phase. The complex is being developed by a joint venture company, Sadara Chemical Company, working with its parents, Dow and Saudi Aramco, Saudi Arabia's national oil and gas company. It will consist of twenty-six manufacturing units occupying about 3 square miles. By the time the complex becomes fully operational in 2016, it will produce more than 3 million tons of product.

Risk Analysis

As discussed in chapter 2, quantitative risk analysis methods are essential tools for enterprise risk management. Risk generally has two dimensions: the *likelihood* that a given outcome will occur, expressed as a probability, and the *magnitude* of the consequences, expressed in terms of financial or other performance metrics. Although the consequences can be either adverse or beneficial, in most cases risk analysis focuses on adverse outcomes that are undesirable for the firm and its stakeholders.

In terms of adverse outcomes, there are generally two classes of risk considered by businesses:

1. *Chronic risks* are associated with slow-moving, cumulative factors (e.g., worker exposures to hazardous conditions) or gradual depletion of critical resources (e.g., groundwater). Such risks are usually assessed in terms of the expected incidence of specific outcomes (e.g., occupational injuries or illnesses, water shortages) over a given time period.

2. *Acute risks* are associated with sudden, episodic events that may have significant consequences, such as hurricanes, transportation

accidents, union strikes, or customer defections. Such risks are commonly assessed in terms of a *risk profile*, which assigns a probability distribution to the range of risk magnitude (see chapter 2).

Risk analysis tools are used for a variety of purposes, including the following:

- *Vulnerability assessment* is the process of identifying and characterizing potential sources of risk (sometimes called hazards or vulnerabilities) in terms of their nature, mechanisms of action, and possible outcomes; this assessment provides a basis for setting priorities to determine which risks merit greater attention.

- *Quantitative risk assessment* is the process of estimating the likelihoods or magnitudes of selected risks that have been identified; it generally includes assessment of the uncertainty associated with the "best" estimates of risk.

- *Integrated risk evaluation* is the process of assigning relative importance to risks that have been identified or quantified, based on regulatory, economic, social, or other factors that influence their acceptability to the enterprise and its stakeholders.

- *Risk management* is the process of deciding how to avoid, mitigate, or otherwise control those risks that are deemed unacceptable; it generally involves a balancing of risks against the costs and benefits of mitigation alternatives.

- *Risk communication* is the process of understanding the concerns of stakeholders regarding identified risks and explaining the results of risk assessment, evaluation, and management decisions in terms that are meaningful to them.

Risk analysis is a complex subject, spanning a broad variety of risk sources, mechanisms, end points, and mathematical techniques. Despite the large amount of literature on risk analysis, the methods are still evolving due both to theoretical advances and new empirical findings. Because there are fundamental limitations on what is knowable, we will continue to rely on predictive models, which cannot be validated except in hindsight. As a consequence, decision making about the mitigation of risks will always be challenged by the presence of significant assumptions and uncertainties in the available information.

For example, health and safety risk analysis requires information about the following:

- Types and magnitudes of risk *agents*, such as hazardous materials, waste and emissions, and ionizing radiation
- Possible initiating *events* leading to unplanned releases, such as leaks, spills, fires, explosions, or deliberate human intrusion
- Fate and transport *mechanisms* that describe how released agents are dispersed in the environment and partitioned among air, water, soil, and other media as well as how they are chemically and physically transformed
- Categories of *receptors* that may be exposed to released agents, including workers, community residents, sensitive populations (children, pregnant women, etc.), natural vegetation and wildlife, aquatic organisms, and domestic animals and crops
- Exposure *pathways*, or routes, whereby humans and other biota may be exposed to released agents or their by-products, including inhalation, uptake through direct contact, ingestion in water, and bioaccumulation in the food chain.

Other categories of business risks that need to be analyzed include price and currency fluctuations, political upheavals, industrial accidents, natural disasters, and stakeholder pressures. Although most of these factors are generally seen as downside risks, they may represent business opportunities. For example, DuPont developed strong internal capabilities in the area of workplace health and safety and then decided to launch a new business that offers these services to other companies. At the same time, DuPont manufactures a variety of safety equipment and specialty materials, such as Kevlar® fibers. The company set a market-facing goal to introduce at least one thousand new products or services by 2015 that help make people safer globally, and then it proceeded to introduce more than twice that number.

Cost-Benefit Analysis

Cost-benefit analysis includes a wide range of methods used to support business decision making. Conventional financial tools such as discounted cash flow analysis or net present value are often used to build a business case by comparing the capital investment and operating costs with the expected flow of returns. Taking a systems approach, however, means that decision makers must go beyond conventional cost accounting methods to consider the broader costs and benefits incurred by the enterprise, its customers, or other parties at various stages of the product life cycle.

Making a business case for resilience investments is particularly challenging because we can only speculate about the types and severity of potential disruptions. If sufficient data are available, techniques such as risk analysis and predictive simulation can be used to explore the possibilities, but in the real world, there will always be significant uncertainties. The best way to justify expenditures on improving resilience is to identify measurable benefits that will be realized even if no disruptions were to occur. For example, investing in alternative energy sources such as geothermal systems can lower the costs of energy supply while improving process reliability (see chapter 10).

Because conventional accounting methods often do not capture the costs or benefits associated with resilience improvements, recognizing these types of business synergies may require some financial creativity. For example, energy costs are often assigned to overhead accounts and cannot easily be traced to a particular business decision. To overcome this problem, some companies have adopted an approach called *activity-based costing*, also known as *total cost assessment*, which involves identification and quantification of direct, indirect, and other costs across the life cycle of a facility, product, or process. This approach is an extension of life cycle costing methods used in the defense sector to manage large, multi-year weapon systems programs in which major costs are associated with deployment, logistical support, and decommissioning. Similar techniques have been used in the computer and other industries to capture the total "cost of ownership" for enterprise assets.

To understand the full scope of costs and benefits, it is helpful to divide them into the following categories:[7]

- **Conventional:** Material, labor, other expenses and revenues that are commonly allocated to a product or process (often called "direct" costs)
- **Potentially hidden:** Costs (or benefits) to the firm that are not typically traced to the responsible products or processes, such as legal fees or safety training courses (often called "indirect" costs)

- **Opportunity**: Costs associated with opportunities that are foregone by not putting the firm's resources to their highest-value use
- **Contingent:** Potential liabilities or benefits that depend on the occurrence of future events, such as potential occupational health and cleanup costs related to a spill of a hazardous substance
- **Goodwill:** Costs (or benefits) related to the subjective perceptions of a firm's stakeholders, such as brand image, customer loyalty, or favorable relationships with regulatory agencies
- **External:** Costs (or benefits) of a company's impacts on the environment and society that do not directly accrue to the business, such as the benefits of reduced waste generation for product consumers

Veolia, the global environmental services company, applied these concepts to develop a tool that measures the "true cost" of water. The tool combines traditional capital and operating costs with estimation of water risks and their financial implications. As illustrated in figure 9.2, it accounts for the direct costs of building and maintaining water infrastructures, the indirect costs such as legal expenses, and unexpected financial burdens during the life of these assets. These cost elements are organized in four categories: *operational* such as water shortages; *financial*, such as commodity price changes; *regulatory*, such as obligation to protect

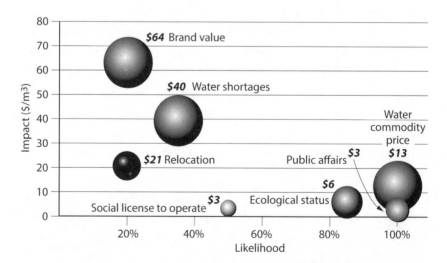

Figure 9.2. Example application of Veolia's true cost of water tool

ecological status; and *reputational*, such as loss of brand value or social license to operate.

Of course, taking a systems approach will complicate the task of identifying all the relevant costs and benefits for a particular business. The first challenge is defining the system boundary. For example, the possible boundaries of analysis for a manufacturing facility could be initial construction costs, operating and maintenance costs, or total value chain costs. The next challenge is identifying who bears the costs. When the scope of decision making extends beyond an individual firm, analysis can become challenging because costs and benefits can be distributed unevenly among business partners, employees, suppliers, customers, and neighboring communities. How to factor in these "externalities" is an ongoing debate among economists and policy makers.

The above managerial accounting approaches take the perspective of a single organization, such as a company or a city government, but to understand financial trade-offs at a community level requires a broader view of public well-being, based on the concept of inclusive wealth.[8] A variety of economic valuation methods are available to approximate the nonmonetary aspects of community sustainability and resilience, including social and environmental costs and benefits. Recent work on economics-based and biophysically-based analyses of ecosystem services has been helpful in classifying these methods.[9]

Economic valuation methods are commonly used to quantify decision trade-offs in monetary terms, based on measures of individual and societal preferences that depend on human values and beliefs. However, traditional valuation approaches are largely static and typically ignore the implications of physical thresholds and nonlinearities. In contrast, noneconomic biophysical approaches, such as life cycle assessment, represent the physical consumption of energy and materials and are useful for potentially identifying physical tipping points or limits, but ignore the value implications of these physical changes in terms of societal costs and benefits.

More recent approaches to nonmarket valuation have been developed in an attempt to value resilience. These approaches integrate economics-based and physically-based methods by modeling the implications of specific changes in terms of human well-being. A nontrivial first step is developing a coupled natural-human systems model to describe the dynamics of the system, such as the 3V model described in chapter 3. The model can then be used to assess the implications of a shift to a less desirable state (e.g., when a critical resource threshold is crossed) by quantifying the gains and losses in human well-being.[10]

Resilience in Action

Crisis Management at IBM[11]

Back in January of 2009, IBM had a wake-up call. Gazprom, the Russian gas monopoly, was embroiled in a pricing dispute with Ukraine. Russian Prime Minister Vladimir Putin, in typical hard-line fashion, decided to cut off natural gas shipments through the Ukraine pipeline, which supplied about 40 percent of Europe's gas imports. The resulting shortages created a hardship for most European nations during a bitterly cold winter and led to rationing of natural gas supplies. This incident foreshadowed the increasing tensions that led to Russia's invasion of Ukraine in 2014.

During this period, IBM had been significantly increasing its activities in growth market countries. The company had established a growth market unit to develop business in rapidly emerging economies, including Brazil, Russia, India, and China (known as the BRIC countries) and other global markets. Rather than managing these businesses centrally, IBM's strategy was to develop a strong in-country presence and hire local managerial talent. While this strategy was sensible, it did involve greater business risks, including acquisition of businesses in countries plagued with military conflicts.

IBM was not substantially affected by the 2009 Ukraine pipeline interruption, but this incident was a warning sign of how unpredictable the international business environment can be. John Paterson, the company's chief procurement officer, was paying attention. He immediately saw the potential vulnerability of IBM or any multinational company to supply chain disruptions caused by geopolitical upheavals or other unforeseen events. Paterson summoned Lou Ferretti, an IBM veteran who had earned a reputation as a problem solver. In the previous decade, Lou had taken charge of preparation for Y2K, contingency planning for pandemics, and other unique challenges. As a supply chain project executive, his broad responsibilities included risk management, environmental compliance, and social responsibility.

In 2009, Paterson gave the challenge to Ferretti to establish a process that would address the cumulative effect of sourcing in growth market countries. This led to the creation of what was eventually called the total risk assessment (TRA) tool and process. Under Ferretti's guidance, a team was organized to design a global database that keeps track of every supplier or service provider in its global "ecosystem," organized by country and supply chain tier. (See figure 9.3.) From a variety of external and internal sources, TRA continuously gathers intelligence about various countries, logistical hubs, suppliers, site locations, commodities, and disruption threats. In total, the TRA tool keeps track of about three thousand supply

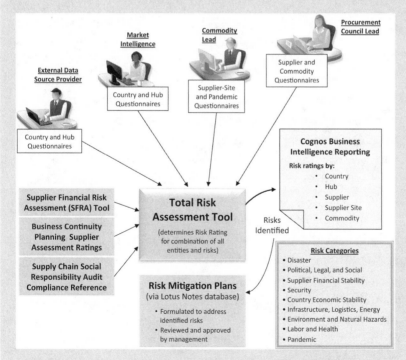

Figure 9.3. *Architecture of IBM's total risk assessment tool*

chains globally. The tool produces risk ratings that can generate automated alerts and also serves as a repository of risk mitigation plans for previously identified risks.

TRA represented a major shift from IBM's previous risk management practices, which were handled separately for different commodities and lacked consistency across risk categories. Given the increased complexity of supply chains and the potential for cascading risks that might be triggered by a minor disruption, IBM's increased use of global sourcing called for a more comprehensive, enterprise-wide means of assessing risk. TRA was fully deployed in 2010, and it was not long before it proved its worth.

In 2011, when the tsunami struck Japan, the resulting widespread power outage disabled many supply chains around the world, but IBM emerged relatively unscathed thanks to TRA. The system was able instantly to identify the affected suppliers in every one of its sixteen commodity groups (memory, logic, etc.) as well as "critical parts," thus enabling IBM to rapidly identify alternative sources to support its purchasing, engineering, and product development functions. IBM did not miss a single shipment nor increase any product's lead time. Before TRA, with each commodity managed by a different organization, this process would have taken several days.

The company established a procurement war room to maintain oversight during the tsunami crisis, and since 2011, the war room has been used on a regular monthly basis for planning exercises; there is usually a crisis brewing in some part of the world that could affect IBM. TRA tracks a broad spectrum of risk issues, including rare earth minerals, power grids, supplier financial health, pandemics, and natural disasters. TRA's algorithm for estimating risk likelihood has been patented, and the tool received an IBM Outstanding Innovation award. It was also a finalist for the 2011 Supply Chain Innovation Award given by the Council of Supply Chain Management Professionals.

IBM has made several strategic moves to render its global supply chains as resilient as possible. Although the company's

manufacturing base continues to shrink in pursuit of greater efficiencies, the company has designed its processes to be completely
interoperable across its worldwide sites. Any product can be manufactured and shipped from any location, providing reserve capacity in the event of supply chain interruptions. In recent years, the
company has made several acquisitions to further strengthen its risk
management technology portfolio, and it is considering how it can
leverage Watson (its iconic supercomputer) to probe "big data" for
global risk characterization.

Driven by a culture of thought leadership from the board of
directors on down, IBM executives are constantly questioning their
practices and considering the megatrends and forces of change that
will challenge their continued success. It is no coincidence that IBM
consistently ranks among the top performers in sustainability. For
example, in *Newsweek*'s 2012 environmental ranking of the top
international and US-based companies, IBM ranked first on the US
list and fourth globally. According to the *Daily Beast*:

> IBM is a poster child for a company evolving with the
> times—first from hardware to software, and now through
> sustainability solutions across energy and water efficiency,
> smarter cities, etc. as categories of opportunity to drive
> future revenue.[12]

Takeaway Points

- Business teams that want to consider resilience need a variety of
  tools for purposes of business innovation, screening and prioritization, decision making, and performance evaluation.
- Qualitative instruments (e.g., SCRAM index) can be applied at
  a business-unit level to assess resilience to potential disruptions,
  identify gaps, and recommend improvement strategies.

- Resilience indicators can be defined for specific types of systems (products, processes, or assets), thus enabling system comparison, monitoring, and adaptive management.

- Key performance indicators can be used to quantify specific aspects of enterprise resilience at a business or enterprise level and to measure progress in resilience improvement.

- When probabilistic forecasting is not possible, scenario-based planning methods can help identify enterprise vulnerabilities anticipate potential disruptions and develop resilience strategies.

- Predictive simulation models based on cause-and-effect logic (e.g., system dynamics) enable companies to project the effects of hypothetical disruptions, including unexpected feedback loops.

- Quantitative methods such as risk analysis and cost-benefit analysis can be helpful for decisions about investing in resilience improvements, provided the necessary data are available.

- **Resilience in Action:** To support its international operations, IBM's procurement organization developed a total risk assessment tool, enabling instantaneous response to global sourcing disruptions.

PART 3

Designing Resilient Systems

CHAPTER TEN

Design for Resilience

*Design thinking starts with divergence, the deliberate attempt
to expand the range of options rather than narrow them.*

Tim Brown[1]

We have looked at many examples of resilient behaviors that enable companies to avoid damaging disruptions and seize opportunities for competitive advantage. Unlike biological systems, companies have the capability to redesign themselves rapidly. We do not have to wait for natural selection to reveal the winners and losers. Companies are self-organizing systems that can anticipate future challenges and deliberately transform their structure and function to increase their resilience. That being the case, it seems only natural that resilience should become part of every company's approach to innovation.

Enterprise Innovation

As mentioned in chapter 5, established companies that are market leaders often have difficulty in recognizing disruptive innovations that challenge their existing business model.[2] Companies that excel at innovation, such as 3M and Apple, spend considerable time and resources establishing disciplined yet flexible internal processes that encourage creativity while systematically selecting and refining the truly worthy concepts. Some companies have explored crowdsourcing of external innovations, whereas others have set up internal "skunkworks" that are protected from the institutional bureaucracy.

To accelerate time to market, most companies use a systematic process for new product development, often called integrated product development or simultaneous engineering.[3] This approach engages cross-functional teams (engineering, manufacturing, marketing, etc.) to ensure early understanding and resolution of key issues that will influence the success of the product, taking into consideration the entire product life cycle: sourcing, manufacturing, distribution, support, maintenance, recycling, and waste disposition. The cross-functional teams work in a parallel, coordinated fashion to address life cycle *requirements*, including quality, manufacturability, reliability, maintainability, safety, and sustainability, which enables them to "get it right the first time" by anticipating performance issues and avoiding costly design changes at a later stage. Codification of this approach has led to design guidelines and tools such as design for manufacture and assembly (DFMA) and design for environment (DFE); these practices are sometimes described as design for X, or DFX.[4] Can resilience be added to the list of requirements?

The answer, not surprisingly, is yes. Resilience can—and should—be an integral consideration for enterprise innovation in the design of products, business processes, facilities, and infrastructure systems. We define *design for resilience* (DFR) as follows:

> The systematic anticipation of disruptive factors that could influence enterprise performance, and the adaptation or transformation of enterprise products, processes, or assets so as to reduce vulnerabilities and improve capabilities, thus enabling sustained or enhanced performance.

Note that DFR must extend beyond the design of products or services. Over the last several decades, the scope of design has evolved from a focus on the artifact (building or product) to an integrated view of the system in which it operates, including broader concerns about unintended environmental consequences such as depletion of scarce resources (see "The Evolution of Design" box). Design for sustainability was the most recent addition to this list of enterprise concerns, embracing not only environmental performance but also questions of economic prosperity, public health and safety, business ethics, labor conditions, human rights, and community well-being. DFR is a further step in that evolution in that it is concerned with the robustness and adaptability of both structural and functional characteristics, including products, processes, buildings, and infrastructure.[5]

The Evolution of Design

In the 1990s decade, businesses were challenged to address unintended environmental, health, or safety impacts over the full product life cycle. The response was design for environment.

In the 2000s, businesses were challenged to address the threat of resource scarcity and diminished opportunity for future generations. The response was design for sustainability.

In the 2010s, businesses are challenged to cope with complexity and turbulent change in global economic and ecological systems. The response is design for resilience.

Control versus Intervention

Among the vast range of systems that we produce and use, there are many well-bounded systems for which the design is truly controllable. Clothing, appliances, aircraft, and buildings, for example, are designed and then manufactured or assembled through a rigorously controlled series of processes. On the other hand, we participate in social and ecological systems that are not at all designed, yet we can still design "interventions": policies and strategies that influence the system behavior and evolution. For example, virtually all our environmental protection efforts are merely interventions: we introduce changes such as soil remediation, pollution reduction, reforestation, or stream flow diversion in an effort to enhance the health of the overall ecosystem. The more we understand nature's designs, the better we can select the appropriate interventions. Sometimes these well-intended interventions can backfire and produce unintended consequences; an example is the introduction of nonnative species that disrupt local ecosystems.

Controlled design is the dominant focus of the product and process development community, and companies have made huge investments in design technology to improve the creativity, quality, and speed of system development. For example, the system acquisition processes used by the US Department of Defense engage dozens of organizations in a controlled design effort, supported by advanced information technology, which encompasses the full life cycle of military systems. Inevitably, such efforts turn out to have unforeseen, inadvertent effects on related systems. For example, the Defense Department recently began to examine the sustainability of domestic artillery test ranges in terms of their impacts on surrounding communities, an issue that was overlooked for many years.

The unforeseen secondary impacts of system design may be trivial or profound. For example, consider automotive design:

- A structural design change that improves vehicle performance may have virtually no significant secondary impacts.

- A change in the materials of construction may have substantial impacts on economic and ecological systems within the manufacturer's supply chain.

- A new engine design may have not only supply chain impacts, but also economic, environmental. and social impacts that affect entire markets.

Examples of Designs that Emphasize Resilience

- **Self-healing materials.** Erik Schlangen, a civil engineer at Delft University in the Netherlands, has developed a technology for self-healing asphalt, thus extending the life of roadways and avoiding the repeated costs of repairing damaged pavement. Ordinary asphalt is mixed with strands of steel wool, which can then be heated by passing an induction coil over the road. The heat melts the sticky bitumen around it, and as the bitumen cools, it mends the asphalt pavement.

- **Reconfigurable computer chips.** There are two types of conventional computer chips: application-specific integrated circuits that are designed to perform a specific computation efficiently and microprocessors that can execute a series of instructions but perform more slowly. An alternative, emerging technology is based on field-programmable gate arrays, which can reconfigure themselves to optimize performance for a variety of different applications.

- **Resilient communication networks.** Modern society and global commerce are highly dependent on the availability of uninterrupted digital communication. Network designers must anticipate a variety of threats, including equipment failures, traffic overloads, malicious attacks, and other disruptions. Accordingly, a broad array of resilience capabilities has been developed; they include fault tolerance, redundancy, diversity, multilayer protection, and cloud backup.

- **Flexible supply chains.** As described in chapter 6, multinational companies that operate global sourcing and delivery networks have designed their supply chains with flexible logistics strategies to accommodate unexpected disruptions. Examples include decentralization of assets, availability of multiple raw material sources and manufacturing locations, and collaboration with customers and suppliers to adjust lead times and schedules.

- **Material recovery networks.** As described in chapters 4 and 7, many manufacturing firms have developed closed-loop solutions for the beneficial reuse of waste materials. One approach, pioneered by the US Business Council for Sustainable Development, is the formation of regional by-product synergy networks in which one company's waste becomes another company's feedstock. Waste recovery not only cuts operating costs, but also hedges against material shortages.

- **Adaptive organizations.** As described in chapter 8, leading companies have learned to cope with a changing business environment by designing organizations that avoid narrow specialization, encourage local autonomy, and emphasize communication across functions. For example, Intel has managed to adapt to successive waves of evolution in the computing industry, thanks largely to its success in nurturing technical innovation and thought leadership.

We cannot design a perfect natural environment or an ideal society, but we can try to modify the controllable characteristics of our designed artifacts (e.g., factories, products) in ways that create environmental and social benefits. As observed by strategy experts Mark Kramer and Michael

Porter, growth and prosperity are linked to the health of the "competitive context," the social and environmental assets that provide employee talent, market demand, and a reliable supply of materials and energy.[6] Any type of product, process, or service innovation can influence these linkages in numerous ways. This concept implies that "design" is more than just creating an artifact; it is a deliberate intervention within a complex set of relationships.

Individual companies, especially market leaders, can exert powerful influences through deliberate system interventions. Sometimes these interventions may have uneven consequences; for example, decisions that improve profits may compromise worker health. Regulatory agencies can also exert powerful influences, but a more promising approach is collaborative partnerships whereby industry, government, and other interested parties jointly design interventions for the benefit of society. Without such partnerships, well-intentioned efforts may go awry; airborne emissions might merely be shifted to waterborne effluents elsewhere in the supply chain, for example, or the distribution of risks and benefits among workers and consumers might be inequitable. Perhaps the most ambitious example of collective system intervention is the worldwide effort to reduce greenhouse gas emissions.

The basic principles of DFR (see "Principles of Design for Resilience" box) suggest that targeted interventions can improve the resilience of the enterprise to unforeseen disruptions and strengthen its position with respect to the network of interdependent systems in which it operates. A specific DFR initiative represents such a purposeful intervention, focused on the design of a product, process, or asset. Note that DFR complements traditional risk management and can support movement toward sustainability (see chapter 12).

Principles of Design for Resilience

- The resilience of human systems, including communities, infrastructures, and enterprises, may be jeopardized by biophysical and socioeconomic constraints or disruptions.

- Human interventions, including new policies, practices, and technologies, can improve the ability of a system to remain in a desired state or enable the system to shift to a preferred state.

- Indicators of relative resilience can be defined for specific categories of similar systems, thus enabling system comparison, monitoring, and adaptive management.

- Human foresight about potential future disruptions can guide the selection of a portfolio of interventions that maintain or strengthen the resilience of managed systems.

- Even in the absence of foresight, it is possible to increase the "inherent" resilience of a system by improving characteristics such as diversity, dispersion, flexibility, redundancy, and buffering.

- Additional information about the probabilities or consequences of specific perturbation scenarios can support the application of risk assessment and management methods.

- Resilience is a necessary but not sufficient condition for achieving sustainability; in particular, there may be trade-offs between short-term resilience and long-term sustainability.

An important design principle for DFR is "inherency": making resilience a natural property of the design rather than an added feature. One example of inherent resilience is the use of distributed architectures. Traditionally, the design of engineered systems, including both hardware and software, has been approached as a process of hierarchical decomposition; that is, the overall system function and architecture are developed first, and then the systems and subsystems are designed accordingly. Such hierarchically organized systems (e.g., aircraft, nuclear plants) tend to have rigid operating parameters; are resistant to stresses or shocks only within narrow boundaries; and may be vulnerable to small, unforeseen perturbations. Alternatively, distributed systems composed of independent yet interactive elements may deliver equivalent or superior functionality with greater resilience. In the electric power field, for example, proponents of alternative energy have argued that distributed, renewable

energy systems are less vulnerable to catastrophic failures than centralized fossil-fuel-based generating sources because of their modularity,
redundancy, and ability to decouple from the grid.[7] Although distributed
systems may not always represent a preferred solution, they represent an
alternative architecture that resembles the patterns seen in living systems.
(See "Inherent Resilience of Distributed Systems" box.)

Inherent Resilience of Distributed Systems

- A collection of distributed electric generators (e.g., fuel cells)
  connected to a power grid may be more reliable and fault-
  tolerant than a central power station.

- A swarm of miniature unmanned surveillance drones may be
  less costly and more robust than a single conventional surveillance aircraft.

- A network of autonomous software agents operating asynchro-
  nously may be more effective and speedier than a monolithic
  software program.

- A geographically dispersed workforce linked by telecommunications may be less vulnerable to catastrophic events that could
  destroy facilities than a centralized workforce housed in a single
  location.

- A decentralized, multi-agent emergency response system may
  be more flexible and dependable than a centralized system that
  incorporates costly fail-safe technologies.

- A global network of business partnerships for a multinational
  enterprise may be more resilient to geopolitical or economic
  upheavals than a series of international acquisitions.

Resilience Enhancement Protocol

Whether a company is designing a new system or improving an existing system, it is useful to have a step-by-step protocol that can guide
cross-functional teams as they consider whether the system is sufficiently

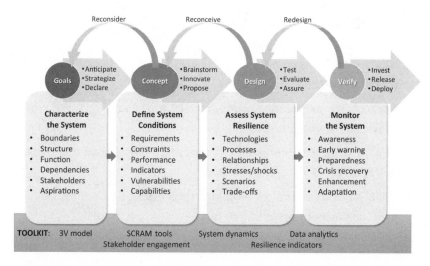

Figure 10.1. *Innovation process with the resilience enhancement protocol*

resilient and design appropriate interventions. Figure 10.1 presents a general protocol based on recent experience in both the private and public sectors that can be incorporated into any innovation process, including product development, capital expansion, and supply chain configuration. The top portion of the figure represents a typical innovation process, which proceeds in four principal stages from goal definition to concept development to design specification to verification. At each stage, the innovation team may discover problems or barriers that require returning to an earlier stage for another iteration before the product can be released.

Supporting the innovation process is an additional layer of investigation that we call the resilience enhancement protocol (REP). The main purposes of REP are to develop an understanding of "affiliated" systems that may affect the performance of the design and to assess the resilience implications. At one extreme, if there are no important interactions anticipated with affiliated systems, the protocol reduces to a standard "stage-gate" process commonly used in industry. At the other extreme, the external interactions may be so significant that the majority of the design effort is spent trying to understand affiliated systems, including impacts and boundary constraints. REP poses the following questions:

- What is our baseline system position regarding key resource flows and interdependencies?

- What are the emerging opportunities for creating value within this broad system?

- What are the emerging threats to availability of the key resources that we will require?

- What are the unintended economic, environmental, or social consequences of our operations?

- What is the nexus among our opportunities, requirements, and external impacts?

- What innovations in technologies, processes, or business models might enhance the system?

- As conditions change in the future, how can we ensure that the system continues to be viable?

Practitioners of REP need to be supported by a toolkit that enables implementation. Examples of such tools, which were discussed in chapter 9, are shown at the bottom of figure 10.1.

Needless to say, this protocol is extremely broad in scope so that in practice it can lead in many directions and differ greatly depending on the context: the type of business, the challenges addressed, the type of innovation or intervention proposed, and the affiliated system characteristics. The four stages of the process—addressing goals, concept, design, and verification—are summarized next.

Strategic Goals: Characterizing the System

The goal definition stage of the innovation process involves anticipating future needs and changing conditions, developing a strategy for addressing those needs, and declaring specific goals for a design initiative or system intervention. One of the greatest challenges at this stage is establishing the scope for a system design effort. As mentioned in chapter 3, every system interacts with its environment and hence is part of a larger system. Carried to the extreme, this type of reasoning leads to the Gaia hypothesis, which claims that the world is a single giant organism.[8] This perspective is not a very helpful if one is trying to design an electronic appliance. A more practical application of systems thinking is to consider the related systems in which a design artifact is embedded—literally thinking outside the box—which may lead to surprising innovations. For example, the development of a telephone answering device might be seen from several alternative system perspectives:

- It is part of a home system from the *consumer use* perspective and can become a network appliance as well a communication device.

- It is part of a material flow system from a *supply chain* perspective and can become a source of reusable components at the end of its life.

- It is an enabling technology from a *developing economy* perspective and can serve as the crucial element of an entrepreneurial communication service business.

It follows that one of the most important steps in system design is establishing a clear, practical definition of the *structure, function,* and *boundaries* of the system. For example, are we creating a voice recording device, a telephone answering system, or a voice mail service? Are we designing a physical device, a human interface, or a total solution delivery process? Depending on the system scope, the design methodology and technological options can vary greatly. For purposes of customer value analysis and life cycle assessment of competing designs, a common way to define systems is in terms of a unit of functional value. In the above example, the functional unit chosen might be a "message capture event" that includes both recording and playback. Then design performance could be analyzed in terms of cost, reliability, ease of use, reliability, and resource efficiency per message capture event.

This stage in the design process is critical because it defines the scope, context, structure, subsystems, linkages, and boundaries of the system to be considered, and explicitly identifies the relevant stakeholder interests. The 3V framework described in chapter 3 can be useful in establishing system boundaries and key stakeholders. Because an overly narrow perspective may ignore important influences or unintended consequences, the definition of goals should address all the important stakeholder groups and aspirations that might be affected by a system intervention. For example, introduction of an alternative energy technology should consider not only its affordability and the expected reduction in greenhouse gas emissions but also the potential for unintended consequences such as displacement of jobs or disruption of community lifestyles. Taking a systems view may also reveal unexpected benefits, such as increased resilience to power failures.

Innovative Concepts: Defining System Conditions

The second stage of the process is developing one or more innovative concepts that promise to meet the strategic goals. A range of product and process designs as well as system interventions can benefit from DFR thinking. Recent advances in genomics, materials science, nanosystems, and information technology can contribute directly to resilience by increasing the efficiency and adaptability of existing products and processes. For example, increased use of electronic communication and virtual meetings reduces the need for more costly physical transportation while enabling businesses to function even when physical infrastructure is disrupted.

A critical step at this stage is the definition of system requirements, which are testable conditions to be met. In traditional systems engineering, requirements have focused on observable product characteristics such as cost, structure (e.g., size, geometry), and functional performance. The conditions may be represented in various forms, ranging from qualitative statements (e.g., "system shall shut down when left idle") to quantitative metrics (e.g., power rating). Testing of requirements typically involves some human interpretation, and the broader the scope of the system, the more difficult it becomes to apply definitive tests. Qualitative assessment tools such as the SCRAM approach discussed in chapters 6 and 9 can be helpful even at the concept stage.

As mentioned in chapter 5, system design may benefit from including requirements that address inherent resilience. Characteristics such as diversity and adaptability may not have an obvious relationship to system performance, but they may contribute to the system's longevity and ultimate success. For example, vehicle designers have increasingly stressed adaptability issues such as reliability and maintainability under extreme conditions, which influence both life cycle performance and cost of ownership. Sometimes the greatest resilience is achieved through design simplicity, which reduces the chances of unexpected failure or disruption.

To incorporate resilience issues, system requirements definition should take into account the boundary conditions derived from affiliated systems. It may require thinking beyond the usual supply chain considerations to consider the broader industrial and social context. A number of multinational companies such as Unilever, Hewlett-Packard, and Procter &

Gamble have become increasingly interested in addressing the large, here-tofore untapped markets at the base of the economic pyramid.[9] They have found that designing products as well as marketing and distribution systems to serve low-income populations in developing nations requires a deep understanding of local conditions, resources, and behavior patterns. For example, soft-drink manufacturers have learned that in a region with severe water shortages, industrial depletion of groundwater should be avoided. Practically speaking, the needs of society must be clarified and prioritized through *stakeholder engagement*, including a dialogue among corporations, government policy makers, and public-interest groups.

The value chain issues that need to be emphasized when defining requirements vary greatly across different industries. Natural resource extraction industries such as mining, agriculture, and forest products might emphasize appropriate land use, ecosystem protection, and worker safety. Industries farther downstream in the supply chain, such as petroleum refining, metals, chemicals, and electric utilities, might emphasize process safety, conversion efficiency, and waste minimization. Industries close to the end customer, such as food and beverage, pharmaceuticals, automotive, and electronics, might emphasize tamper-proof packaging (for consumable products), end-of-life recovery (for durable products), and responsiveness to social needs. Finally, service industries such as transportation, communication, and retailing might emphasize efficient logistics, product certification (e.g., lumber), and human resource development.

Design: Assessing System Resilience

The third stage of the process is development of detailed designs corresponding to the innovative concept; included here is testing these designs to ensure that they meet the stated requirements. This stage typically involves definition of performance indicators, evaluation of alternative solutions relative to the current baseline, and consideration of trade-offs and synergies. In this stage, the process shifts from conceptual analysis to quantitative assessment and evaluation of the expected costs, risks, and benefits for various stakeholder groups. The choice of appropriate indicators is critical for assessing trade-offs among different alternatives, and it is important to analyze uncertainties as well as sensitivity of the results to key assumptions. Chapter 9 discussed the selection of resilience indicators to be included in the set of performance indicators. For example, Cisco is using its resiliency index to evaluate new product introductions so that design teams can assess choices about supply chain partners and components. This kind of assessment allows Cisco to build supply chain resilience into the design of the product rather than trying to de-risk the supply chain after the product launch.[10]

At a system level, there are often hidden trade-offs to consider. Systems that are smaller, faster, cheaper, and more flexible may actually be less resilient in terms of manufacturing needs than their bulkier, slower predecessors. In the semiconductor industry, for example, the precision manufacture of a tiny microchip consumes large amounts of energy and materials and requires highly controlled, clean room conditions.[11] Nanosystem production may prove even more resource-intensive, and some scientists are concerned about the potential hazards associated with inhalation of nanoparticles. Due to system-level considerations, no technology can be deemed intrinsically resilient or sustainable. Renewable or bio-based materials are not necessarily preferable to inorganic materials. Recycled materials are not necessarily preferable to virgin materials. Biodegradable materials are not necessarily preferable to durable materials because by-product synergy networks may recycle spent materials into new applications. It all depends on the system boundaries and requirements.

The system design process itself is undergoing considerable evolution. Traditional, hierarchical design has proved cumbersome in the sense that a single deviation can disrupt the entire process. Many organizations are experimenting with new techniques such as cooperative, distributed, asynchronous design. Again, this approach is patterned after the self-organizing behavior of living systems. Through advanced communication and groupware technologies, design teams can be distributed geographically and share their ideas and progress via interactive computer displays. For design of complex systems, the ability to iterate rapidly is especially important because design teams need to assess the robustness of alternative designs under a variety of different scenarios and assumptions. In the automotive industry, for example, it is common for the major automakers to colocate design engineers from their principal first-tier suppliers together with their in-house design teams, enabling a tightly integrated development process.

A key part of any iterative system development effort is evaluating the anticipated performance of a partially or fully completed design. In many industries, this evaluation process is supported by automated tools, known as computer-aided design and manufacture, which can perform highly detailed simulations before the system is ever built. Boeing, for example, has honed this approach to the point that its engineers can design a complete aircraft without ever physically building and testing a proto-type. As the design focus shifts from form and function to the impact on affiliated systems, however, the evaluation task becomes immensely more challenging. As described in chapter 9, it is possible to develop high-level simulations to evaluate the anticipated effects of interventions on urban and regional ecosystems or socioeconomic systems, but we can seldom ensure their validity.

Verification: Monitoring System Deployment

The final step in the process is development of a plan for the release, introduction, and deployment of the system. In the past, product development teams often released their specifications to the manufacturing organization without considering downstream issues associated with component procurement, product distribution, customer support, maintenance, waste disposition, and potential upgrades of the original design. The introduction of simultaneous engineering methods and more rigorous stage-gate reviews has helped shift these considerations back into the design process, thus avoiding delays and unnecessary costs. The question of broader system impacts is still poorly understood, though, and it is rare for design teams to consider them explicitly.

Companies wishing to pursue resilient and sustainable system design need to consider the broad implications of an innovative system on all the enterprise stakeholders. Each stakeholder group will be touched by the system in different ways and will have particular expectations. For example, employees expect the system to be safe and easy to operate. Shareholders expect the system to streamline existing operations and provide an improved return on an investment. Customers expect efficacy and convenience. Public-interest organizations expect environmental and social benefits. To understand and manage all these expectations, it is critical for the company to engage with its stakeholders, understand their concerns, and develop mutual trust so that the system can be introduced successfully. Indeed, the design team should begin considering the deployment phase as early as possible, including establishment of data-collection mechanisms to enable analysis of outcomes.

The importance of stakeholder engagement is evident in the life sciences field, where companies are developing a host of biotechnology-based products that they claim will enable a shift to sustainable agriculture. One controversial issue in this field is the introduction of genetically engineered pest-resistant crops. Proponents have claimed that this technology will reduce pesticide use, increase agricultural productivity, and lower consumer costs, whereas opponents are concerned about unforeseen health and environmental impacts and long-term resilience of agricultural assets such as biodiversity and soil quality.[12] Thus, history has shown that it is wise for designers to consider not only the direct benefits of a technological innovation, but also the socioeconomic system into which it will be introduced.

Finally, companies must be prepared to adapt to unexpected consequences by modifying or redesigning products, processes, or systems. Frequent product recalls, as often seen in the automotive industry, are undesirable, but it is important to protect the brand and restore customer confidence by responding quickly to hidden flaws. Companies must build capabilities to respond to any sort of crisis, even if they are not at fault. That was evident in the swift response of Johnson & Johnson to the Tylenol poisoning incident in 1981, which ultimately led to new standards for protective packaging on consumer products.

Takeaway Points

- Enterprise innovation should include design for resilience to ensure the inherent fitness of products, processes, and facilities for a turbulent business environment.

- The scope of design is broadening from a focus on the artifact to an integrated view of the value chain and the broader economic, environmental, and social consequences of design choices.

- Design for resilience can identify interventions that enhance the capacity of the enterprise to anticipate potential disruptions, reconfigure its assets to cope with extreme events, and adapt to change.

- Design interventions may include new policies, procedures, or technological innovations that make a product, process, or asset more resilient, taking into account cost and performance trade-offs.

- The typical stage-gate innovation process can be augmented by a resilience enhancement protocol that considers the broader system characteristics and potential disruptions.

CHAPTER ELEVEN

Connecting with Broader Systems

If we see each problem—be it water shortages, climate change, or poverty—as separate, and approach each sepa-rately, the solutions we come up with will be short-term, often opportunistic quick fixes that do nothing to address deeper imbalances.

Peter Senge[1]

In chapter 5, we saw that enterprise resilience extends well beyond product development, touching on virtually every business process from supply chain management to capital investment. Through resilience awareness, companies are able to harness the unique capabilities of human beings to reflect on experience, interpret new information, recognize systemic patterns, envision alternative futures, avoid unexpected consequences, and creatively adapt to unforeseen situations. These capabilities can be leveraged through the practice of DFR: examining every decision through a resilience lens and developing guidelines for building inherent resilience into the structure and function of all enterprise systems.

Because the success of an enterprise depends on external systems—nature and society—the DFR approach can include interventions in those systems as well. The interconnections between industrial and human systems, from family scale to national scale, are important to understand: cultural forces can either enhance or obstruct enterprise resilience. Likewise, the interconnections between economic and ecological systems provide us with access to natural capital, including valuable ecosystem services that we have barely begun to understand and use. Traditional design protocols are based on static assumptions about these external systems, but DFR requires investigation of changing boundary conditions and potential disruptions in interconnected systems.

Scope of Design for Resilience

The traditional focus of risk management has been on protecting tangible enterprise assets, although business continuity management extends to intangible enterprise assets and even community assets such as the transportation infrastructure. The traditional focus of design has been on tangible objects such as products and buildings, although "sustainable design" practices extend more broadly to services, business processes, and supply chains. Likewise, the scope of DFR should be extremely broad. Possible DFR interventions could range from designing a building so that it can withstand natural disasters to collaborating with a university on curriculum design so that future employees will have critical decision-making skills. There are literally no boundaries.

Generally, we can define three categories of assets for which resilience can be strengthened (figure 11.1):

1. **Enterprise assets** include all assets owned, managed, or controlled by the enterprise as well as important structural and functional characteristics such as intellectual capital.

2. **Community assets** include assets and characteristics of stakeholder communities that are affiliated with the enterprise due to geographic proximity or socioeconomic linkages. Such communities may include customers, suppliers, employees, investors, regulators, municipalities, regions, advocacy groups, professional organizations, or industry associations.

3. **Shared assets** include products and services that are transferred between the enterprise and affiliated communities as well as assets and characteristics that are jointly owned, managed, controlled, or otherwise influenced.

These three categories are each divided into *tangible* assets that have concrete physical form or monetary value and *intangible* assets that are abstract, although they can often be monetized.

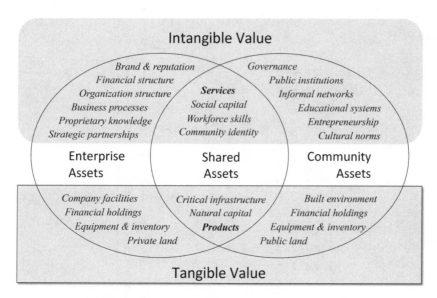

Intangible Value

Brand & reputation   Governance
Financial structure   Public institutions
Organization structure   **Services**   Informal networks
Business processes   Social capital   Educational systems
Proprietary knowledge   Workforce skills   Entrepreneurship
Strategic partnerships   Community identity   Cultural norms

Enterprise   Shared   Community
Assets   Assets   Assets

Company facilities   Critical infrastructure   Built environment
Financial holdings   Natural capital   Financial holdings
Equipment & inventory   **Products**   Equipment & inventory
Private land   Public land

Tangible Value

Figure 11.1. Potential targets of design for resilience interventions

Based on the systems approach introduced in chapter 3, we can strengthen enterprise resilience by considering not only innovations in product and service systems, but also potential interventions in overall enterprise systems, community systems, ecological systems, and infrastructure systems. Table 11.1 shows how each of these different types of systems can be characterized according to the four principal resilience attributes defined in chapter 5: adaptability, cohesion, efficiency, and diversity, or ACED.

When considering enterprise *structure*, these ACED attributes apply along several different dimensions:

- **Physical structure and location of enterprise assets, including facilities and equipment:** Resilience depends on configuration, connectivity, redundancy, modularity, and buffer capacity. An example is the existence of multiple manufacturing plants serving different regional markets.

- **Organizational structure of departments and human resources, both formal and informal:** Resilience depends on diversity, versatility, accountability, and internal communication channels. An example is multifunctional training that enables employees to assume different responsibilities in the event of a business disruption.

Table 11.1. Resilience attributes of various types of systems

	Adaptability	Cohesion	Efficiency	Diversity
Product or service system	Customer-friendly customization; failure recovery	Strong brand identity; unique or distinctive features	Value delivered relative to total expenditures	Platform with many possible extensions or configurations
Enterprise system	Organizational learning; cash reserves	Unifying corporate culture; strong partnerships	Agile decision processes; resource productivity	Encouragement of diverse business strategies
Community system	Transparency and flexibility of major institutions	Geographic boundaries; strong cultural identity	Affordable means for human needs satisfaction	Ethnic, cultural, institutional, and political variety
Ecological system	Tolerance and assimilation of pollution	Natural habitat boundaries; tightly clustered food web	Robust ecological cycling of energy and nutrients	Biodiversity (i.e., large variety of plant and animal species)
Infrastructure system	Fault detection and mitigation; reserve capacity	Region-wide coordination of planning and control	Avoidance of waste or leakage; emphasis on asset utilization	Multiple resources, delivery channels, and technologies

- **Financial structure of business units, subsidiaries, and joint ventures:** Resilience depends on solvency, flexibility, insurance coverage, and availability of emergency funds. For example, cooperative banks such as Rabobank consist of a network of semiautonomous member banks with a distributed governance structure. These banks have displayed greater resilience during financial crises than investor-owned banks.[2]

- **External structures, including social, economic, and ecological linkages at multiple scales, such as geographic, temporal, and institutional:** Enterprise resilience depends on stakeholder trust and credibility, collaborative relationships, access to resources, and broader system resilience. Key stakeholder groups include employees, suppliers, contractors, customers, investors, regulators, communities, and advocacy groups. An example is a public-private partnership to foster climate resilience and adaptation strategies, such as the BICEP coalition mentioned in chapter 1.

Similarly, when considering enterprise *function*, the ACED attributes may apply along the following dimensions:

- **Physical processes, including material acquisition, handling, manufacturing, transportation, and facility maintenance:** Resilience depends on awareness and rigor in conduct of these processes. An example is the alertness of purchasing managers to anomalies in supplier behavior.

- **Organizational processes, including strategic and tactical decision making in major line and staff functions:** Resilience depends on foresight, adaptability, experimentation, and learning. An example is demand forecasting using data analytics to interpret early warning signals of market fluctuations.

- **Behavioral processes and characteristics at the individual and team level:** Resilience depends on resourcefulness, trust, motivation, persistence, and adaptive capacity under stress. An example is voluntary participation of employees as first responders during a crisis.

- **External social, economic, and ecological processes, including natural resource use, market and consumer behavior, political or social movements, and global conflicts:** Resilience depends on engagement, awareness, creativity, preparedness, crisis management, and opportunity identification. Examples that were mentioned in chapter 4 include waste elimination through industrial ecology and fostering shared value with upstream supplier communities.

The following sections describe resilience considerations in urban communities and energy systems. These two particular types of external systems have vital interdependencies with every enterprise.

Resilience in Urban Communities

Cities are perhaps the most complex and turbulent of all human systems, yet they remain extraordinarily resilient. Like living organisms, cities have survived, adapted, and flourished through the centuries, overlaying different cultures, lifestyles, and technologies in a rich and evolving mosaic. Today, cities are a crucible of change, where social, economic, and environmental pressures are intensified and where the challenges of resilience and sustainability converge.

Due to steady migration away from rural areas and traditional lifestyles, more than half of the planet's inhabitants now live in cities. Dozens of "megacities" support more than 20 million inhabitants, where wealth flourishes alongside poverty, crime, and despair and where infrastructure systems are severely stressed. In the United States, some cities have achieved revitalization, whereas others are plagued by urban decay and a flight to the suburbs. What all cities share are two basic challenges: balancing economic prosperity with quality of life (i.e., sustainability) and overcoming disruptions that threaten human safety or business continuity (i.e., resilience).

Many promising urban initiatives have emerged, including smart growth, waste-to-energy conversion, greener buildings, and vertical farming.[3] Innovative companies are entering this space and discovering new markets. One example is IBM's worldwide "smarter cities" initiative, described later in this chapter. In addition, cities can serve as living laboratories to test innovative technologies or policies aimed at improving health, education, neighborhood stability, economic vitality, security, and safety.

Federal agencies, including the Department of Housing and Urban Development, the Department of Homeland Security, and the Department of Commerce are also investigating community vulnerabilities and resilience improvement strategies. As mentioned in chapter 7, the federal government recognizes that national security is no longer merely concerned with defense of US interests against hostile attacks; it also includes protection of our sources of food, energy, water, and materials, which are the foundation of community prosperity. In response to mounting evidence of climate change, a 2014 presidential task force issued broad recommendations for the federal government to help communities by removing barriers to investment in resilience, modernizing federal grant and loan programs, and developing information and tools for climate change preparedness and adaptation.[4]

In particular, many cities are concerned about the "nexus" described in chapter 4 that connects water, energy, and food. Dwindling water resources threaten to disrupt energy and food production, while rising energy prices threaten to increase the costs of supplying both food and water. Moreover, all three of these critical resources depend on the availability of land, materials, and infrastructures. Recent droughts in the western United States have highlighted our precarious dependence on water: we take it for granted until it is suddenly unavailable. One outstanding exception is New York City, which established a system that has ensured water security for many decades (see "Water Resilience in New York City" box).

A string of urban infrastructure crises, from collapsing bridges to floods to pipeline explosions, has illustrated the hidden vulnerabilities that we face today. Pursuing business as usual has become riskier in the face of increasing stresses and shocks. One recent report proposed that governments can take a more adaptive approach called anticipatory governance.[5] This approach will require improving foresight in the face of uncertainty, coordination of governance bodies to develop cohesive policies, and monitoring of consequences for purposes of adaptive management.

Water Resilience in New York City

An example of resilience in water resource management is the New York City water supply, widely considered to be the gold standard of urban water systems. The system draws on a large watershed, where rainwater and snowmelt make a three-month journey through gravity-powered aqueducts stretching more than 100 miles from the Catskills to the edges of New York City. It covers parts of eight New York counties and a sliver of Connecticut, and includes nineteen reservoirs and three controlled lakes, with a total capacity of approximately 580 billion gallons of water supplied to about ten million people. The system was constructed in the 1940s, 1950s, and 1960s as low-lying upstate farms and small towns were flooded to create reservoirs. From these reservoirs, the water is channeled into underground aqueducts that are considered engineering marvels. The water is then held in smaller reservoirs and is finally distributed through huge pipes that feed the five boroughs of New York City. The 90-acre Hillview Reservoir is one of the last stops for the water before it reaches the city; it holds about 1 billion gallons, roughly equivalent to one day's worth of water demand in New York City.

The National Academy of Sciences has underscored the need to build resilience in US communities, including flexibility, adaptive capacity, and infrastructure redundancy. One recent study[6] recommends that the federal government incorporate national resilience as a guiding principle. The report suggests that flexible risk management strategies are needed, involving multiple stakeholders and a mix of structural improvements and policy tools. To justify investments in resilience, communities will need assurance that there will be measurable benefits, including improved prosperity and quality of life even in the absence of a disaster. This thinking underscores the need for resilience indicators to assess issues such as infrastructure performance, building integrity, and social and business capacity for disaster recovery.

The Rockefeller Foundation established the 100 Resilient Cities program to develop collective wisdom about resilience strategies (see chapter 5). It has found that no matter what their size or location, resilient cities seem to share certain core capabilities: constant learning, rebounding rapidly from shocks, limiting the effects of failure, adapting flexibly to change, and maintaining spare capacity. True resilience is not just about responding to disasters, but also dealing with stresses such as unemployment, urban violence, and food or water shortages. One of the most important lessons emerging from this program is that resilient cities are able to turn tragedy into opportunity, rebuilding to become stronger than before. Included in this process is an awareness of the environmental and social factors that enable a city to remain healthy, vibrant, and diverse.

Resilience in Energy Systems

Energy systems are critical to the functioning of both enterprises and communities. In the United States and other developed countries, companies have grown to depend on a reliable supply of fuel, natural gas, electricity, and steam to power their factories and supply chains. When interruptions do occur, most companies have backup energy systems to continue critical operations, but business continuity can be hampered by power failures or other disasters that affect the regions in which they operate.

Today, there are significant vulnerabilities due to the interdependencies among energy and other infrastructure systems, which can lead to cascading failures. For example, Hurricane Katrina in 2005 caused power outages in New Orleans, which in turn led to contamination of the city water supply and loss of phone service. Outages can also interrupt the availability of transportation fuels because most pumps and compressors rely on electric power.

At the same time, there is increasing pressure on the utility industry to shift to cleaner and more efficient sources of power, such as natural gas and renewable fuels, in an effort to reduce greenhouse gas emissions and avoid depletion of fossil-fuel reserves. One advantage of distributed energy technologies such as fuel cells, windmills, and solar panels—in addition to improving environmental sustainability—is that they increase resilience in the event of central power interruptions.

A variety of emerging technologies will help improve both the resilience and sustainability of energy systems.[7] For example, the introduction of "smart grid" technology will enable rapid detection of outages as well as modularization and decoupling of local neighborhood power networks. Unfortunately, current standards require that grid-connected solar systems shut down automatically during a power failure, underscoring the need for better energy storage technologies.

There is general agreement among resilience practitioners that we need to improve coordination, awareness, and planning on the part of state and local governments as well as partnerships with the business community. The importance of foresight was made apparent by Superstorm Sandy, which crippled the New York/New Jersey region for weeks. Afterward, the State of New York commissioned a blue-chip study that came up with comprehensive recommendations for strengthening the resilience of energy and other infrastructure systems to minimize the impacts of future disasters (see "Energy Resilience Recommendations" box).

Energy Resilience Recommendations[8]

Strengthen critical energy infrastructure

- Facilitate process of securing critical systems.
- Protect and selectively place underground key electrical transmission and distribution lines.
- Strengthen marine terminals and relocate key fuel-related infrastructure to higher elevations.
- Reinforce pipelines and electrical supply to critical fuel infrastructure.
- Waterproof and improve pump-out ability of steam tunnels.
- Create a long-term capital stock of critical utility equipment.

Accelerate the modernization of the electrical system and improve flexibility

- Redesign the electric grid to be more flexible, dynamic, and responsive.
- Increase distributed generation statewide.
- Make the grid electric vehicle ready.

Diversify fuel supply, reduce demand for energy, and create redundancies

- Facilitate greater investments in energy efficiency and renewable energy.

- Diversify fuels in the transportation sector.
- Support alternative fuels across all sectors.
- Lower the greenhouse gas emissions cap (if available).

Develop long-term career training and a skilled energy workforce

- Create a workforce development center.
- Expand career training and placement programs.
- Build awareness of the need for skilled workers.
- Coordinate workforce development among all stakeholders within the energy sector.

An important part of the resilience enhancement protocol, described in chapter 10, is for companies to engage with stakeholders, including governmental agencies and utility companies, to strengthen these infrastructure capabilities. In particular, companies such as Johnson Controls, Siemens, and General Electric have an important role to play in sharing their expertise with stakeholder groups and responding to emergencies. For example, in the days after the 2011 Japanese tsunami, General Electric was able to dispatch temporary truck-mounted gas turbines that served as immediate, portable power sources. Without functioning backup diesel generators, portable power was crucial to emergency responders struggling to control damage and to customers otherwise left in the dark.

Although continuity of energy supply is important, a more innovative approach to resilience is reducing or altogether eliminating dependence on energy supply. There has been growing interest in energy efficiency though the development of green buildings, ambient-temperature processes, combined heat and power, and other technological improvements. Even more radical is the movement toward zero-energy operations, exemplified by the US Army's Net Zero program, which has set goals for its military bases to achieve self-sufficiency in terms of energy and water use as well as zero waste generation. The motivations are twofold: greater operating efficiency and associated cost reductions are an obvious benefit, but more importantly, Net Zero provides resilience against interruptions in utility services and thus ensures mission readiness in times of crisis. Dozens of Army bases around the United States have already made strides toward net zero energy by adopting renewable technologies such as geothermal and solar energy.[9] The resulting reductions in greenhouse gas emissions are a welcome side benefit.

Resilience in Action

IBM and Smarter Cities[10]

One of the most effective marketing campaigns in recent decades was based on the Smarter Planet concept introduced by IBM in 2009. It was actually more than a marketing campaign; it was a way of envisioning a better world based on digital technology and underscored IBM's role as an agent of change. Since its introduction, it has generated many spin-off concepts and helped earn IBM hundreds of millions of dollars in contracts.

The Smarter Planet concept grew out of several exploratory initiatives. One was called Big Green Innovation, echoing the "Big Blue" nickname that became synonymous with IBM's traditional mainframe business. Big Green originated from an online "Innovation Jam" in 2005, sponsored by IBM's CEO, Sam Palmisano. He posed a simple question to the IBM workforce: "What ideas will matter for IBM and the world?" Out of the 140,000 replies received, the ten best ideas were selected, and Palmisano funded their development with $10 million apiece.

IBM was already doing work on energy and power management, and a cluster of new ideas emerged around the theme of the environment, including issues such as water and waste management and sustainable agriculture. These ideas were combined to form Big Green, a startup environmental business unit established in 2006 with Peter Williams as the chief technology officer.

The Smarter Planet concept was actually a convergence of ideas from two different IBM groups. One was the Smarter Planet marketing team, led by John Iwata and John Kennedy, which was creating "smarter" value propositions for different business sectors.

The other was a think-tank team within the IBM Academy that was working on a concept called "instrumented planet." The latter group, including Williams, was exploring the disruptive business implications of having a vast, interconnected network of intelligent sensors and meters appearing everywhere, providing a massive stream of data that could be mined for useful information. Today, this network is commonly described as the Internet of Things.

As these groups engaged with each other and reached out to various customers, it became clear that there was significant overlap. In 2011, Smarter Cities was formed as a thread of Smarter Planet that focused on the needs of urban areas. Big Green was absorbed into this initiative, and Williams remained involved with Smarter Cities.

Smarter Cities addresses the following question: How can we use information to enhance how cities operate, including operations and broader communications? For example, sensor technology can help sense environmental changes rapidly and provide early warnings. The proliferation of social networks and mobile computing can provide a sort of connective tissue that enables community cohesion, especially in times of stress. IBM's consulting and technology solutions for water and energy management are clearly relevant to the challenge of urban resilience.

Over the years, IBM has developed and commercialized a broad range of information technology applications that contribute to urban resilience, including weather forecasting, flood modeling, and structural monitoring for levees. In 2004, for example, IBM deployed a disaster response team to help with recovery from the Indian Ocean tsunami, including the creation of survivor databases. The tools IBM developed for this purpose were donated to the Sahana Foundation and are now in the public domain. These experiences led to a growing realization that IBM is actually in the resilience business.

In 2012, Williams began to work with Dale Sands of AECOM on detection systems and response to earthquake hazards. IBM and AECOM, together with Willis Re, serve on the Private Sector Advisory Group of the United Nations International Strategy for

Disaster Reduction (UNISDR), which developed the "ten essentials" of disaster resilience in 2005 as part of the Hyogo Framework (see "UNISDR's Ten Essentials for Making Cities Resilient" box). IBM and AECOM jointly developed a resilience scorecard based on these ten essentials. Cities can use this self-assessment tool to evaluate their preparedness, including collaboration, risk assessment, building codes, natural buffers, and warning systems.[11]

From these innovative exploratory steps, a business strategy for IBM began to take shape: it would become the "essential" company for resilience in the face of both gradual stresses and acute shocks. As cities, companies, and agencies concerned with resilience gathered more and more information, IBM would provide valuable services and technology to collect, analyze, integrate, and display the information, enabling it to be used effectively.

The Smarter Cities team adopted the concept of an intelligent operations center (IOC), which underpins several IBM capabilities in the water, transportation, and public safety arenas. Williams describes IOC as a "situation room on steroids." It includes geographic information systems, optimization tools, workflow management, and a dashboard for real-time situation monitoring. The team further developed this concept through partnerships with several cities around the world, including Manila, Rio de Janeiro, and Rotterdam. With the increasing focus on disaster resilience, the IOC evolved into an emergency management center that incorporates a critical asset management tool called Maxima.

One important application of the IOC was in Rotterdam, which sits on a delta formed by the Rhine, Meuse, and Scheldt Rivers and is one of the world's largest and busiest ports. For centuries, the Netherlands has developed methods to defend its low-lying coastline against flooding from the North Sea. In fact, approximately 70 percent of Netherlands land area and 55 percent of housing are below sea level. After catastrophic flooding occurred in 1953, claiming more than 1,800 lives, Rotterdam erected a massive system of levees and flood barriers. With rising sea level, increasing storm intensity, and land subsidence, however, it became apparent that these defenses were not sufficient and that a more integrated approach was needed.

Accordingly, a program of innovation called Flood Control 2015 was launched by a consortium of Dutch companies, research institutes, and engineering consultants. The program improved flood protection by ensuring the timely availability of critical information: even a few hours can make a major difference in disaster preparedness, prevention, and damage mitigation. Rather than focusing strictly on defensive barriers, the program introduced new strategies, such as making room for the river by means of floodplains. A cornerstone of the program was the use of advanced forecasting and decision-support systems that take advantage of innovative sensors to provide a holistic view of all the important environmental factors.

As part of this program, IBM teamed up with the City of Rotterdam and others to develop the Digital Delta, a state-of-the-art flood monitoring and management system based on the IOC architecture. This system has several pioneering elements:

- An integrated national network of air, water, and soil monitoring stations captures thousands of data points daily on hundreds of variables such as temperature, quality, salinity, wave heights, speed, direction, pressure, and clouds.

- Smart levees augment satellite observation with "Geobeads" (geotechnical sensor strings produced by Alert Solutions BV), enabling real-time continuous infrastructure monitoring and integrity modeling to provide up to forty-eight hours advance warning of levee instability.

- Advanced urban water cycle management avoids sewer overflow and flooding by achieving end-to-end integration of precipitation, infiltration, runoff, and storage data; for example, underground parking garages can be used as overflow reservoirs.

- An integrated water distribution system accounts for social and economic needs, using multiobjective optimization based on large amounts of data from heterogeneous sensors and enabling collaboration between national, regional, and local water managers.

- Citizen engagement in crisis response is enabled by a system called Crisis Buzz that performs automated data mining of social media streams and supplies validated information to citizens and crisis response teams, dramatically reducing response time.

Ever seeking growth through innovation, IBM has expanded its Smarter Cities portfolio to connect with several existing business areas, including Smart Grid and Smart Buildings. Sales to private-sector organizations as well as cities are yielding significant shareholder value while reinforcing IBM's reputation as a thought leader in the use of information technology.

UNISDR's Ten Essentials for Making Cities Resilient[12]

1. Put in place organization and coordination to understand and reduce disaster risk, based on participation of citizen groups and civil society. Build local alliances. Ensure that all departments understand their role in disaster risk reduction and preparedness.

2. Assign a budget for disaster risk reduction and provide incentives for homeowners, low-income families, communities, businesses, and the public sector to invest in reducing the risks they face.

3. Maintain up-to-date data on hazards and vulnerabilities, prepare risk assessments, and use them as the basis for urban development plans and decisions. Ensure that this information and the plans for your city's resilience are readily available to the public and fully discussed with them.

4. Invest in and maintain critical infrastructure that reduces risk, such as flood drainage, adjusted where needed to cope with climate change.

5. Assess the safety of all schools and health facilities and upgrade these as necessary.

6. Apply and enforce realistic, risk-compliant building regulations and land use planning principles. Identify safe land for low-income citizens and upgrade informal settlements, wherever feasible.

7. Ensure that education programs and training on disaster risk reduction are in place in schools and local communities.

8. Protect ecosystems and natural buffers to mitigate floods, storm surges, and other hazards to which your city may be vulnerable. Adapt to climate change by building on good risk-reduction practices.

9. Install early warning systems and emergency management capacities in your city and hold regular public preparedness drills.

10. After any disaster, ensure that the needs of the affected population are placed at the center of reconstruction, with support for them and their community organizations to design and help implement responses, including rebuilding homes and livelihoods.

Takeaway Points

- Enterprise resilience is dependent on the resilience of connected systems, including natural capital, human capital, and social capital, especially when external conditions are changing.

- The design for resilience approach should extend beyond internal processes to consider the broader systems in which the design is embedded: buildings, infrastructure, communities, economic sectors, and ecosystems.

- Design for resilience interventions can help strengthen important assets that are shared with or controlled by stakeholder communities affiliated with the enterprise.

- The resilience of urban systems can be strengthened by threat awareness, response coordination, and improved adaptability and recovery capacity, as well as consideration of longer-term system performance.

- The resilience of energy systems can be strengthened by diversifying the energy mix and providing flexibility, buffer capacity, and the ability to decouple in the event of power failures.

- **Resilience in Action:** IBM's Smarter Cities initiative has discovered strategic business opportunities in helping cities improve their resilience to disasters.

CHAPTER TWELVE

Looking Ahead: From Resilience to Sustainability

The Internet is becoming the town square for the global village of tomorrow.

Bill Gates[1]

W e live in a crowded world of ever-increasing connectivity, with both cooperation and conflict occurring on a global scale. Individuals, companies, and communities are linked through worldwide systems of communication, transportation, and commerce. Similarly, individual products and services are linked to the global value chains in which they are created, delivered, and used. As we have seen, this connectivity presents daunting challenges to the design and deployment of new products, processes, and assets. Instead of focusing purely on the function and form of a product or service, design teams today must consider a broad range of system-level issues, including safety, security, manufacturability, serviceability, material and energy efficiency, end-of-life recovery, environmental emissions, and long-term impacts on quality of life for future generations.

When we introduce the resilience perspective, a new challenge emerges: how to cope with global forces that can disrupt the functioning of an enterprise, a supply chain, a community, or an entire region. In the face of such complexity, traditional methods for analyzing costs, benefits, and risks can become overwhelming. Instead, it is helpful to delve more deeply into the fundamental properties of successful systems in the biological, social, and commercial arenas. What do such systems have in common? They exhibit adaptive, self-organizing behavior that enables continuity in response to external stresses and shocks. They can survive unexpected disruptions, although they can also fail catastrophically. When we design new systems, we cannot anticipate all future possibilities, but we can endow them with intrinsic characteristics that improve their resilience: adaptability, cohesion, efficiency, and diversity. To succeed in the new global village, enterprises must learn to embrace change and apply systems thinking.

This book has provided a variety of tools, guidelines, and case studies to help enterprises practice design for resilience, but at least one more puzzle remains to be solved. How can an enterprise balance the long-term, idealistic goals of sustainability with the immediate, practical goals of resilience? Are sustainability and resilience synergistic, or are there inherent conflicts? Some would argue that sustainability includes resilience as a necessary condition for coping with the unexpected and flourishing in the face of change. Others would argue that resilience includes sustainability as a necessary condition for operating in balance with environmental constraints and social expectations. Of course, everyone is right. To avoid semantic confusion, we need to delve more deeply into these concepts. An enterprise cannot have separate strategies for sustainability and resilience because they are deeply intertwined.

Sustainability Challenges

The need for a transition to a sustainable economy is becoming ever more urgent.[2] The productive capacity of the planet is already stressed in meeting current demand for energy, goods, and services while billions of people remain mired in poverty, lacking even basic hygiene. According to the Millennium Ecosystem Assessment, global ecosystems are severely degraded,[3] and many believe that we have already overshot the planet's ecological capacity.[4] Responding to these warning signals, various sustainability principles have been proposed by organizations such as the Coalition for Environmentally Responsible Economies,[5] the United Nations Environment Program,[6] and the Natural Step.[7] These principles share many common elements, including waste elimination, natural resource protection, and equity assurance for present and future generations.

In industries ranging from resource extraction (e.g., petroleum, lumber) to conversion and processing (e.g., chemicals, electric power) to consumer goods (e.g., packaged foods, electronics), shareholders and analysts alike have become sensitized to a company's ecological and social footprint, including global issues such as climate change and poverty. Many leading corporations have adopted a commitment to sustainability, recognizing that environmental protection and social responsibility are important to both shareholders and other stakeholders, including employees, customers, investors, communities, regulators, business partners, advocacy groups, and other nongovernmental organizations. Despite their heightened awareness and commitment, however, most companies have found it difficult to translate broad goals and policies into day-to-day decision making. Progress has been incremental, and global environmental threats such as climate change, soil erosion, and depletion of natural resources have not abated.

Several practical barriers limit the application of sustainability principles. For one thing, the concept of sustainability is quite abstract. The typical manager or employee has trouble understanding the connection between avoiding child labor and promoting biodiversity, let alone relating these issues to his or her own job. The notion of protecting future generations seems quite remote in the face of contemporary business pressures. Society as a whole is slow to respond to these long-term concerns; even climate change is not fully accepted despite clear evidence of a palpable threat. How much polar ice needs to melt away before we finally get it?

Another important barrier is strategic relevance. Sustainability is often associated with resource constraints and maintenance of the status quo rather than with opportunities for continued innovation, growth, and prosperity. The popular metaphor of the triple bottom line[8] seems to imply that profitability needs to be balanced against environmental and social benefits, whereas in truth these three aspects of corporate performance are inseparable and contribute synergistically to shareholder value.[9] Everyone knows there is really only one bottom line, and many fear that expenditures on sustainability will have a negative effect on profits. Although it has been shown that economic, environmental, and social progress can be mutually reinforcing, the business case for sustainability rests heavily on enhancing intangible value drivers rather than directly generating cash.

Alternative definitions of sustainability abound. For design purposes, however, we offer the following definition: "A product, process, or service contributes to sustainability if it constrains environmental resource consumption and waste generation to an acceptable[10] level, supports the satisfaction of important human needs, and provides enduring economic value to the business enterprise."[11]

Note that a product cannot be "sustainable" in an absolute sense; rather, it must be considered in the context of the supply chain, the market, and the natural environment. Therefore, the key practical challenge of sustainable design is to understand how products, processes, and services interact with these broader systems. The 3V framework shows that full exploration of this question extends beyond the enterprise, even beyond the conventional product life cycle, to the underlying resource flows that link economic, social, and natural capital.

An example of this challenge is evident in the quest for what is called sustainable mobility, defined by the World Business Council for Sustainable Development as the ability for humans to "move freely, gain access, communicate, trade and establish relationships without sacrificing other essential human or ecological values, today or in the future."[12] A key issue is how future transportation technologies and demand patterns will evolve, together with their supporting infrastructures (e.g., adoption of hydrogen fuel cells for automotive vehicles will require development of a new refueling network). The infrastructure question is especially challenging because it encompasses roads, railways, airports, and intermodal freight terminals as well as maintenance and guidance systems. Ultimately, the evolution of mobility systems will be influenced by urban and regional planning policies as well as emerging technologies such as driverless vehicles. The vast scope of these interlocking systems is bewildering for any business enterprise seeking to develop a sustainable business strategy for mobility-related products and services.

Rethinking Sustainability

The forces that threaten sustainability, including global warming, ecosystem degradation, and poverty in developing nations, cannot be addressed adequately with one-dimensional solutions that focus on specific improvements, such as energy efficiency. The 3V framework makes it clear that social, environmental, and economic systems are interconnected. Economists point to a paradox called the rebound effect in which decreasing waste will increase economic efficiency, which results in more goods and services being consumed, which in turn causes a net increase in pollution and waste. For example, advances in technology have lowered the cost of lighting, communication, and computing so that we can afford many more electronic gadgets in our homes and businesses. All these devices have become more efficient, but our energy use keeps climbing.

Given the hard reality of finite planetary resources, ecological economists such as Herman Daly have argued for shifting away from a growth-oriented paradigm toward the concept of a steady-state economy, with limits on physical throughputs and stocks.[13] Living systems need to grow, however. So far, real economic growth has been an inevitable consequence of maintaining free and competitive markets. From a purely political perspective, eliminating growth would seem to be a nonstarter.

Some futurists paint optimistic scenarios of a cooperative, harmonious global economy, with advanced technologies enabling efficient utilization of resources.[14] The Rocky Mountain Institute claims that investing in energy efficiency and renewable sources can eliminate fossil-fuel use for electricity; vastly reduce demand for liquid fuels; generate $5 trillion of economic value; and enhance US competitiveness, resilience, and security.[15] Similarly, McKinsey has projected that improvements in resource productivity can lead to a more prosperous and sustainable economy.[16]

The inescapable truth is that neither companies nor policy makers can predict, let alone control, the course of human affairs. Idealized scenarios of a sustainable world in which material and energy cycles become perfectly balanced seem to be a distant fantasy. As the world grows hyperconnected and the rate of change accelerates, the future becomes increasingly obscure. Humans have created order on an unprecedented scale, giving us the illusion of control, but we are more vulnerable today than ever. (Our confidence may have been shaken a bit by recent natural disasters.) The type of order that we create is different from nature's order; it is more tightly coupled, more rigid, and more brittle. The inevitable waves of change will eventually disrupt even the most elaborate structure.

Nature has the solution to excessive growth: natural selection. Enterprises and ecosystems are living systems and follow similar patterns, except that human foresight and intervention enable more rapid adaptation. Companies that are unable to handle the increasing complexity, connectivity, and uncertainty of the global economy will become vulnerable to disruptions and will not survive. Those that adopt sustainability and resilience will prevail. With planning and foresight, however imperfect, we can learn from natural systems and design industrial systems that are better fit for the journey.

My colleagues and I have argued that we need to improve the inherent resilience of our public and private institutions so that society is better equipped to deal with short-term discontinuities (such as hurricanes) as well as long-term stresses (such as carbon emissions). Our recommendation is to foster purposeful collaboration between business and government: "It is essential to anticipate change, understand early warning signals, and take steps to avoid, reduce, and mitigate future problems. A new, more systemic approach to problem solving is needed to avoid unintended consequences, anticipate alternative future scenarios, and strengthen resilience in the face of uncertainty."[17]

Understanding the dynamic relationships among human and natural systems will help planners develop more-resilient strategies that reduce vulnerability to unforeseen catastrophes, enable continued growth, and respect ecological resource capacity. In short, we can design for resilience.

How Resilience and Sustainability Are Coupled

Generally speaking, sustainability and resilience are distinct but mutually reinforcing. For example, reducing the supply chain footprint can help insulate companies from the pressures of climate volatility. As shown in figure 12.1, the more sustainable the system, the better its fitness to flourish because it is less vulnerable to resource shortages or other hardships that may arise from unpredictable disruptions.[18] Conversely, the more resilient the system, the greater its continuity because it is less likely it is to suffer setbacks that would compromise progress toward sustainability goals. Therefore, sustainability improvement initiatives should consider the resilience of both human and ecological systems, including their capacity to adapt to changing conditions (e.g., commodity prices, precipitation rates), and buffer against unexpected disruptions (e.g., power failures, terrorist attacks).

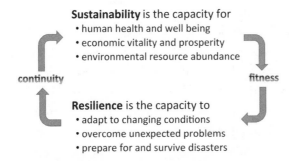

Sustainability is the capacity for
- human health and well being
- economic vitality and prosperity
- environmental resource abundance

continuity

fitness

Resilience is the capacity to
- adapt to changing conditions
- overcome unexpected problems
- prepare for and survive disasters

Figure 12.1. Sustainability and resilience are mutually reinforcing

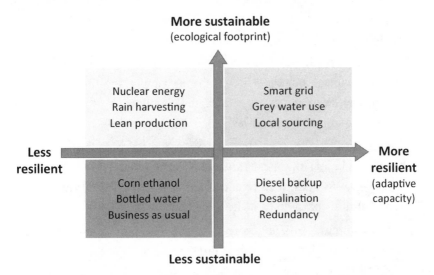

More sustainable
(ecological footprint)

Nuclear energy
Rain harvesting
Lean production

Smart grid
Grey water use
Local sourcing

Less resilient

More resilient
(adaptive capacity)

Corn ethanol
Bottled water
Business as usual

Diesel backup
Desalination
Redundancy

Less sustainable

Figure 12.2. Examples of synergies and trade-offs between sustainability and resilience

There can, however, be trade-offs between sustainability and resilience, as illustrated in figure 12.2. Some technologies and business practices are neither sustainable nor resilient; for example, corn ethanol provides an inferior return on energy and competes for agricultural resources that are critical to food security.[19] Other energy technologies, such as smart grid, hold the promise of both increased efficiency and improved recoverability through distributed generation.[20] Rainwater harvesting is an appealing sustainability practice, but it is vulnerable to droughts. Likewise, leaner production methods may reduce waste, but achieving resilience typically requires investment in reserve capacity.

We can overcome many of the barriers to sustainability by using a new language that is relevant to business interests rather than relying on stakeholder pressures and moral arguments. Resilience thinking provides an immediate and intuitive motivation for businesses to improve their competitive advantage. Building resilience will simply expand the breadth of possible future conditions under which the enterprise remains viable.

Resilience thinking means viewing the enterprise as a living system that is closely coupled with a variety of social, environmental, and economic systems. An enterprise that aspires to sustainability can begin the journey by enhancing its own resilience relative to the systems in which it operates. In particular, it can strive for resilience along the three classic dimensions of sustainability:

1. **Economic resilience** reflects the financial prosperity and stability of the enterprise, including the economic vitality and diversity of the communities in which it operates, the supply chains that it rests on, and the markets that it serves.

2. **Social resilience** reflects the human and social capital of the enterprise, including the capability, teamwork, and loyalty of its workforce; the strength of its relationships and alliances; and the political and cultural cohesion of its host societies.

3. **Environmental resilience** reflects the operational efficiency and effectiveness of the enterprise in terms of resource use and waste minimization as well as its ability to protect and nurture the natural ecosystems in which it operates.

The path forward for creating a sustainable and resilient enterprise will require a new mind-set that embraces constant change, adaptation, and innovation. Building a resilient organization is no easy task; it requires balancing in the zone between adaptability and efficiency, between diversity and cohesion, and between order and chaos. We can learn a lot from mimicking ecosystems, which have developed resilient characteristics over eons of experimentation. We can also learn from other cultures, especially developing nations that have been able to find leapfrog solutions to the challenges of a chaotic environment. To achieve true sustainability, however, we must combine the inherent survival ethic of resilience with a sense of fairness and social responsibility. Table 12.1 provides some basic

Table 12.1. Points to consider for actions affecting sustainability and resilience

Sustainability Questions	Resilience Questions
1. Will the action protect human health and the environment? Will it integrate and optimize environmental, economic, and social benefits?	1. Does the action take into account the full spectrum of risks and disruptive forces that may affect both human and natural well-being?
2. Will the action conserve natural resources—energy, water, materials, land, ecosystems, and air—through prudent use or reuse, protection, or restoration?	2. Does the action recognize the interdependence of the built environment, infrastructure, and natural systems, including the potential for cascading failures?
3. Does the action reflect an orientation toward life cycle thinking, multimedia pollution prevention, minimizing wastes and toxics, and advancing multiple goals through a systems approach?	3. Will the action help to reduce the exposure and vulnerability of critical industrial and ecological assets to extreme events, such as natural disasters or catastrophic failures?
4. Does the action consider the full diversity of available policy and program tools to stimulate and reinforce sustainable outcomes, innovating and collaborating wherever necessary?	4. Does the action increase the inherent robustness, reliability, flexibility, agility, or effectiveness of existing economic and social activities, even for unforeseen threats?
5. Will the action improve people's lives and create healthier communities rather than just correcting problems? Does it consider vulnerable groups (e.g., children, elderly), who may bear disproportionate burdens?	5. Does the action take advantage of a diverse portfolio of resilience capabilities, which may be available from both public- and private-sector organizations that share common goals?
6. Does the action identify meaningful sustainability outcomes and include appropriate metrics? Are there plans to track progress, learn from experience, and adjust strategies accordingly?	6. Does the action seek to strengthen the resilience of existing systems by learning from prior disruptions, innovating, and adapting rather than simply returning to "normal" operation?
7. Does the action include plans to share as much information as possible and engage citizens to take active responsibility for achieving sustainable outcomes?	7. Will the action identify leading indicators of potential disruptions, keep track of external forces and trends, and identify new scenarios that may create future challenges?

Source: Adapted from A. D. Hecht and J. Fiksel, "Solving the Problems We Face: The U.S. Environmental Protection Agency, Sustainability, and the Challenges of the 21st Century," *Sustainability: Science, Practice, and Policy* 11, no. 1 (Spring 2015).

questions that every decision maker should consider when taking actions that may affect the sustainability or resilience of an enterprise, a community, or a broad geographic region.

Toward a Sustainable Future

This book has made the case that resilience is a fundamental property of living systems, which are best understood in terms of dynamic, multilayered networks. Resilience is essentially the ability of a system to resist disorder and maintain a dynamic equilibrium. In a turbulent, real-world environment, resilient systems are able to survive and grow by adapting successfully to unforeseen changes. A business enterprise is also a living system, constantly striving to sense and respond to emerging vulnerabilities or opportunities and sometimes adapting to external stresses by transforming its structure or function.

We often hear shrill cries that the planet is in peril. Doomsayers present a bleak picture of humanity's future prospects, but humans are compassionate, ingenious, and powerful when aroused. Our problem is not helplessness; it is complacency. In any case, it is our self-indulgent lifestyle that is in peril. The planet will survive, with or without human intrusion. Perhaps a more positive approach is to learn how we can live in harmony with planetary ecosystems.

Sustainability is often misinterpreted as a goal to which we should collectively aspire. In fact, sustainability is not a reachable end state; rather, it is a fundamental characteristic of a dynamic, evolving system. Long-term sustainability will result not from movement along a smooth trajectory, but rather from continuous adaptation to changing conditions. We cannot assume that nature will be infinitely resilient, nor can we presume to foretell what cycles of change will occur in the future. A sustainable society must be based on a dynamic worldview in which growth and transformation are inevitable. In such a world, innovation and adaptation will enable human societies—and enterprises—to flourish in harmony with the environment.

As systems grow larger and more structured, their resilience can wane, making them vulnerable to external disruptions and internal decay. The nonlinear nature of complex systems implies that circumstances can change suddenly and that surprises are inevitable. A resilience mind-set involves embracing variability rather than struggling to maintain constancy. Instead of resisting deviations from a "normal" state, resilient organizations recognize early signals of change and respond swiftly to maintain their performance and continuity. At the same time, their planning horizon must be long enough to consider the trade-offs between short-term gains and long-term outcomes.

Managers can use a variety of strategies to enhance the resilience of their particular enterprise components or processes. The path toward improving enterprise resilience involves the following kinds of initiatives:

- Understanding the existing network of enterprise linkages
- Performing baseline assessments of enterprise resilience, including assets, workforce, and product line perspectives
- Developing a strategy for leveraging resilience to drive shareholder value
- Incorporating resilience indicators as a measure of intangible value creation
- Anticipating the full scope of disruptive risks and opportunities
- Supplementing risk management with exploratory thinking and resilience enhancement
- Including resilience requirements in the design of new products and processes as well as capital investment decisions
- Expanding the classic continuous improvement mantra of "plan-do-check-act" to a new cyclical approach: "sense, respond, learn, adapt, design, evolve"

This book has provided but a glimpse of the challenges ahead as system design moves from the bounded, controllable scope of traditional products and services to the boundaryless, unpredictable realm of industrial, ecological, and social systems. The increasing connectedness of these systems creates new opportunities but also exposes society to greater risks. Economic threats such as the collapse of markets, political threats such as military aggression, biological threats such as mutant viruses, or ecological threats such as global warming have become the concern of all nations. Moreover, system complexity keeps increasing: serious proposals are being raised for development of sustainable systems on a much larger scale, such as entire cities or regions.[21]

A systems approach reveals how enterprises and communities are linked to the environment and how they can flourish in harmony with natural systems. We are beginning to understand the resilience of these systems and to study their cyclical patterns of growth, collapse, and renewal. Traditional modeling and forecasting tools, however, are only valid in small regions of time and space where conditions remain relatively constant. Complex, nonlinear systems cannot be modeled by linking together a fragmented collection of linear models. Research is necessary to develop realistic, dynamic models of resilient systems, enabling us to better prepare for extreme disruptions. We need a new generation of analytic tools that incorporate connectivity and turbulent change as fundamental themes rather than afterthoughts. Building realistic models of resilient systems will enable us to prepare for surprises and design for the unforeseen.

Einstein reputedly said, "Individuality is an illusion created by skin." Indeed, separateness is a convenient assumption that enables the analysis of objects, people, or companies as if they were independent of their surroundings. Obviously, this assumption is invalid. System design should proceed with a constant awareness of related systems, boundary conditions, external effects, and potential feedback loops. As design teams continually expand the system boundary, they will need to address new technical challenges in creative ways. For example:

- Requirements will include system behaviors rather than just outcomes.
- Predictive modeling will give way to exploratory scenario-building.
- Design strategies will rely on intervention rather than control.
- Robustness will be achieved through resilience rather than resistance.
- Risk management will draw on new concepts like adaptability and diversity.
- Key performance indicators will be include fundamental resilience attributes.

The resilience perspective has important implications for companies that wish to become more sustainable than they currently are. It is not sufficient for a company to redesign only those systems that it fully controls. At best, that approach will result in incremental changes that do no harm but that do not create substantial benefits either for the enterprise or for society. Instead, companies that wish to ensure their long-term resilience must reach beyond their own boundaries, develop an understanding of the intricate systems in which they participate, and strive for continuous innovation and renewal. In this broader playing field, the rules are different. Strategic adaptation becomes more important than strategic planning, and decision makers need to embrace uncertainty rather than trying to eliminate it.

Finally, it is important to understand the limitations of resilience thinking:

- Resilience is essentially an amoral concept; it is entirely possible for highly resilient systems (e.g., dictatorships) to violate core human values. One needs only to witness the extraordinary resilience of terrorist groups such as Al Qaeda and criminal organizations such as the Mafia. The primary motivation for survival and growth must be supplemented by a commitment to justice and human rights.

- Resilience is typically utilitarian in the pursuit of persistence and performance. It preserves the system function and identity but does not necessarily consider whether the system has a transcendent purpose such as creating value for society. Without a sense of purpose, a resilient enterprise would be hollow, lacking in fundamental motivation and inspiration for its employees and stakeholders.

The history of technological progress has emphasized the conquest of nature, using brute force and standardization to overcome nature's infinite diversity.[22] Today, scientists and engineers are learning from nature, discovering patterns that they can apply for the benefit of both humans and the environment. In business as in science, the old Newtonian view of an orderly, machine-like world is giving way to a new view of a chaotic, evolving world. Designing systems that are inherently resilient will support our collective quest for sustainability in this ever-changing, unpredictable universe.

Takeaway Points

- In a hyperconnected, rapidly changing global village, enterprises need to expand their system boundaries and consider a range of external factors that drive both long-term sustainability and short-term resilience.

- Progress toward sustainability is hampered by its broad scope, its distant time horizon, and perceptions that it conflicts with the fundamental growth objectives of the enterprise.

- Resilience and sustainability are mutually reinforcing: a sustainable enterprise is better fit to overcome costly disruptions than other enterprises, and a resilient enterprise is better able to maintain continuity toward strategic goals.

- There are trade-offs between sustainability and resilience because reducing the ecological footprint may also reduce adaptive capacity; some technologies, however, are both sustainable and resilient.

- One of the first steps for an enterprise on the journey to sustainability is to enhance the economic, social, and environmental resilience of the systems with which it is mutually dependent.

- To achieve long-term sustainability, companies must gain a greater awareness of interdependent dynamic systems, multiple stakeholders, and time scales as well as the limitations of their sphere of control.

Notes

Chapter 1

1   P. Drucker, *Managing in Turbulent Times*, HarperCollins, New York, 2008.

2   J. Fiksel, "Sustainability and Resilience: Toward a Systems Approach," *Sustainability: Science, Practice, and Policy* 2, no. 2 (2006): 14–21.

3   A. Yu. *Creating the Digital Future*. Free Press, 1998, p. 93.

4   T. Kuczinski and K. Irwin, *Severe Weather in North America*, Munich Re, 2012.

5   N. N. Taleb, *The Black Swan: The Impact of the Highly Improbable*, Random House, New York, 2010.

6   J. Fiksel, *Design for Environment: A Guide to Sustainable Product Development*, 2nd ed., McGraw-Hill, 2009.

7   Global Footprint Network, April 12, 2013, http://www.footprintnetwork.org/en/index.php

8   G. Hamel and L. Välikangas, "The Quest for Resilience," *Harvard Business Review*, September 2003.

9   R. Starr, J. Newfrock, and M. Delurey, "Enterprise Resilience: Managing Risk in the Networked Economy," *strategy+business* 30 (Spring 2002).

10 J. Fiksel, "Sustainability and Resilience: Toward a Systems Approach," *IEEE Management Review* 35, no. 3 (2007): 5–15.

11 P. Evans, "Resilient, Sustainable Infrastructure: Shifting Strategic Landscape," presentation at the Future of Energy, Bloomberg New Energy Finance Summit, New York, April 23, 2013.

12 J. A. Nickerson and T. R. Zenger, "Being Efficiently Fickle: A Dynamic Theory of Organizational Choice," *Organization Science*, September/October 2002: 547–566.

Chapter 2

1   N. N. Taleb, *The Black Swan: The Impact of the Highly Improbable*, 2nd ed., Random House, 2010, p. xxxii.

2   Technically, figure 2.1 represents a continuous probability distribution, and the total area under the curve must be 1. A similar chart can be created for a discrete distribution with a finite number of possible outcomes.

3   M. Buchanan, *Nexus: Small Worlds and the Groundbreaking Science of Networks*, Norton, 2002.

4   Committee of Sponsoring Organizations of the Treadway Commission (COSO), *Enterprise Risk Management: Integrated Framework*, AIPCA, 2004.

5   National Research Council, *Science and Decisions: Advancing Risk Assessment*, National Academy of Sciences Press, 2009.

6   G. Dickinson, "Enterprise Risk Management: Its Origins and Conceptual Foundation," *Geneva Papers on Risk and Insurance* 26, no. 3 (2001): 360–366.

7   B. Herbane, D. Elliott, and E. M. Swartz, "Business Continuity Management: Time for a Strategic Role?," *Long Range Planning* 37, no. 5 (2004): 435–457.

8   R. Zolkos and M. Bradford, "BP Disaster Caused by Series of Risk Management Failures, According to Federal Investigation of Gulf Spill," *Business Insurance*, September 18, 2011, http://www.businessinsurance.com/article/20110918/NEWS06/309189982

9   H. Kunreuther, "Risk and Reaction: Dealing with Interdependencies," *Harvard International Review* 28, no. 3 (Fall 2006): 37–42.

10   National Research Council, *Sustainability and the U.S. EPA*, National Academies Press, September 15, 2011.

11   http://www.weforum.org/issues/global-risks

12   World Economic Forum, *Global Risks 2014*, 9th ed., Geneva, 2014, p. 11.

13   N. N. Taleb, *Antifragile: Things That Gain from Disorder*, Random House, 2012.

14   Taleb, *Antifragile*, p. 14.

15   Helpful information was provided by David Bresch of Swiss Re and Brent Dorsey of Entergy.

16   http://www.entergy.com/content/our_community/environment/GulfCoastAdaptation/Building_a_Resilient_Gulf_Coast.pdf

17   A more holistic assessment of the value at risk from climate change would ideally include second-order macroeconomic impacts, human lives, and socioeconomic factors such as human health and ecosystem degradation.

Chapter 3

1   D. Meadows, *Thinking in Systems*, Chelsea Green, 2008, p. 169.

2   D. J. Watts, *Six Degrees: The Science of a Connected Age*, Norton, 2003, p. 303.

3   P. Senge, *The Fifth Discipline*, Doubleday, 1990, p. 3.

4   F. Heylighen, C. Joslyn, and V. Turchin, eds., *Principia Cybernetica*, 2000, http://pespmc1.vub.ac.be/

5   M. Iansiti and R. Levien, *The Keystone Advantage: What the New Dynamics of Business Ecosystems Mean for Strategy, Innovation, and Sustainability*, Harvard Business Review Press, 2004.

6   S. Hart and M. Milstein, "Global Sustainability and the Creative Destruction of Industries," *Sloan Management Review* 41, no. 1 (Fall 1999).

7   R. Starr, J. Newfrock, and M. Delurey, "Enterprise Resilience: Managing Risk in the Networked Economy," *strategy+business* 30 (Spring 2002).

8   J. Fiksel, "A Systems View of Sustainability: The Triple Value Model," *Environmental Development*, June 2012.

9   J. Fiksel, "A Framework for Sustainable Materials Management," *Journal of Materials*, August 2006: 15–22.

10   P. Hawken, A. Lovins, and L. H. Lovins, *Natural Capitalism: Creating the Next Industrial Revolution*, Rocky Mountain Institute, 2001.

11   H. C. Binswanger and R. N. Chakraborty, "Economics of Resource Management," University of St. Gallen, Institute for Economy and the Environment, October 2000, paper commissioned by the European Commission.

12   J. Fiksel, J. Low, and J. Thomas, "Linking Sustainability to Shareholder Value," *Environmental Management*, June 2004.

13   J. D. Sterman, *Business Dynamics—Systems Thinking and Modeling for a Complex World*, McGraw-Hill, 2000.

14   http://ohioenergyresources.com

15   A. Bassi and J. Fiksel, "An Economic Analysis of Ohio's Renewable and Energy Efficiency Standards," *Technological Forecasting and Social Change*, forthcoming.

16   J. Fiksel, R, Bruins, A. Gatchett, A. Gilliland, and M. ten Brink, "The Triple Value Model: A Systems Approach to Sustainable Solutions," *Clean Technology and Environmental Policy*, June 2014.

Chapter 4

1   B. Walker and D. Salt, *Resilience Practice: Building Capacity to Absorb Disturbance and Maintain Function*, Island Press, 2012, p. 24.
2   J. Benyus, *Biomimicry: Innovation Inspired by Nature*, William Morrow, 1997.
3   E. Cimren, J. Fiksel, M. E. Posner, and K. Sikdar, "Material Flow Optimization in By-product Synergy Networks," *Journal of Industrial Ecology* 15, no. 2 (April 2010): 315–332.
4   J. Venetoulis, D. Chazan, and C. Gaudet, *Ecological Footprint of Nations*, Redefining Progress, 2004.
5   K. S. McCann, "The Diversity-Stability Debate," *Nature* 405 (2000): 228–233.
6   M. Csikszentmihalyi, *Flow: The Psychology of Optimal Experience*, Harper and Row, 1990.
7   R. Foster and S. Kaplan, *Creative Destruction: Why Companies That Are Built to Last Underperform the Market—and How to Successfully Transform Them*, Doubleday, 2001.
8   C. Folke, S. R. Carpenter, B. Walker, M. Scheffer, T. Chapin, and J. Rockström, "Resilience Thinking: Integrating Resilience, Adaptability and Transformability," *Ecology and Society* 15, no. 4 (2010): 20.
9   E. Schrodinger, *What is Life?*, Dublin Institute for Advanced Studies, 1943.
10   L. Gunderson and L. Protchard Jr., *Resilience and the Behavior of Large-Scale Systems*, Island Press, 2002.
11   A. deGeus, *The Living Company*, Harvard Business School Press, 1997.
12   N. Nohria, W. Joyce, and B. Roberson, "What Really Works," *Harvard Business Review*, July 2003.
13   E. Werner and R. S. Smith, *Journeys from Childhood to Midlife: Risk, Resilience, and Recovery*, Cornell University Press, 2001.
14   L. H. Gunderson and C. S. Holling, eds., *Panarchy*, Island Press, 2002.
15   S. Hart and M. Milstein, "Global Sustainability and the Creative Destruction of Industries," *Sloan Management Review* 41, no. 1 (Fall 1999).
16   National Intelligence Council, *Global Trends 2030: Alternative Worlds*, National Intelligence Council, 2012.
17   J. Fiksel, "Evaluating Supply Chain Sustainability," *Chemical Engineering Progress* 106, no. 5 (May 2010): 28–38.

Chapter 5

1   G. Hamel and L. Välikangas, "The Quest for Resilience," *Harvard Business Review*, September 2003, p. 63.
2   R. Kupers, ed., *Turbulence: A Corporate Perspective on Collaborating for Resilience*, Amsterdam University Press, 2014, p. 9.
3   Y. Sheffi, *The Resilient Enterprise*, MIT Press, 2005.
4   J. Fiksel, J. Low, and J. Thomas, "Linking Sustainability to Shareholder Value," *Environmental Management*, June 2004.
5   J. Low and P. Kalafut, *Invisible Advantage: How Intangibles Are Driving Business Performance*, Perseus Press, 2002.
6   M. Porter and M. Kramer, "Creating Shared Value: How to Reinvent Capitalism—and Unleash a Wave of Innovation and Growth," *Harvard Business Review*, January/February 2011.

7   J. Fiksel, "Designing Resilient, Sustainable Systems," *Environmental Science and Technology*, December 2003: 5330–5339.

8   H. Lee, "The Triple-A Supply Chain," *Harvard Business Review*, October 2004.

9   C. Folke, S. R. Carpenter, B. Walker, M. Scheffer, T. Chapin, and J. Rockstrom, "Resilience Thinking: Integrating Resilience, Adaptability and Transformability," *Ecology and Society* 15, no. 4 (2010): 20.

10 C. M. Christensen, *The Innovator's Dilemma: When New Technologies Cause Great Firms to Fail*, Harvard Business School Press, 1997.

11 *Watson Wyatt's Human Capital Index: Human Capital as a Lead Indicator of Shareholder Value*, 2001, http://www.blindspot.ca/PDFs/HumanCapitalIndex.pdf

Chapter 6

1   This chapter is partly based on the following published article: J. Fiksel, M. Polyviou, K. L. Croxton, and T. Pettit, "From Risk to Resilience," *Sloan Management Review*, Winter 2015.

2   World Economic Forum, "New Models for Addressing Supply Chain and Transport Risk," World Economic Forum, 2012.

3   World Economic Forum, 2012.

4   K. B. Hendricks and V. R. Singhal, "The Effect of Supply Chain Glitches on Shareholder Wealth," *Journal of Operations Management* 21, no. 5 (2003): 501–522.

5   K. Jacobs, "Hurricane Sandy Hits Travel, Cargo: Costs Unclear," October 29, 2012, www.reuters.com

6   K. Bhasin, "Holiday Shopping Is Being Threatened by Crippled Supply Chains," November 5, 2012, www.businessinsider.com

7   Deutche Post DHL Group, "A High-Performance Network Even during a Crisis," November 5, 2010, http://www.dpdhl.com/en/media_relations/abonnements/finan cial_media_newsletter/background_q1_2010.html

8   C. Burritt, "Walgreen Drops Cardinal Health for AmerisourceBergen Deal," March 19, 2013, http://www.bloomberg.com/news/articles/2013-03-19/walgreen-alliance-boots-win-right-for-amerisourcebergen-stake

9   R. Mason-Jones, B. Naylor, and D. R. Towill, "Lean, Agile or Leagile? Matching Your Supply Chain to the Marketplace," *International Journal of Production Research* 38, no. 17 (2000): 4061–4070.

10 M. Christopher and M. Holweg, " 'Supply Chain 2.0': Managing Supply Chains in the Era of Turbulence," *International Journal of Physical Distribution and Logistics Management* 41, no. 1 (2011): 70.

11 "Resilience—Surviving the Unthinkable," *Logistics Manager*, March 1, 2004: 16–18.

12 Federal Reserve Board, "The Beige Book" (April 13, 2011), http://www.federalreserve.gov

13 UN Global Compact and Accenture, "CEO Study on Sustainability 2013: Architects of a Better World," 2013.

14 Accenture, "Reducing Risk and Driving Business Value: Carbon Disclosure Project," 2013.

15 M. E. Webber, "Will Drought Cause the Next Blackout?," *New York Times*, July 24, 2012.

16 http://blogs.wsj.com/digits/2010/05/24/fake-bp-twitter-account-draws-followers-with-oil-spill-satire/

17 T. J. Pettit, J. Fiksel, and K. L. Croxton, "Ensuring Supply Chain Resilience: Development of a Conceptual Framework," *Journal of Business Logistics* 31, no. 1 (2010): 1–22.

18 Pettit, Fiksel, and Croxton, "Ensuring Supply Chain Resilience," p. 6.

19 Pettit, Fiksel, and Croxton, "Ensuring Supply Chain Resilience," p. 6.

20 Additional information about SCRAM tools can be obtained from the Center for Resilience, Ohio State University. See http://resilience.osu.edu/CFR-site/scram.htm.

21 J. McIntyre and S. Hemmelgarn, "How One Business Made Its Supply Chain More Resilient" (presentation at the Annual Global Conference of the Council of Supply Chain Management Professionals for the 2011 Supply Chain Innovation Award, Philadelphia, October 4, 2011).

22 Helpful information was provided by the following individuals at L Brands: Rick Jackson, Mark Crone, Thomas Hellman, John Joseph, Chris Robeson, and Mike Sherman. The case study was coauthored by Iliana Filyanova of Ohio State University's Fisher College of Business.

23 National Association of Manufacturers and the National Retail Federation, "The National Impact of a West Coast Port Stoppage," Inforum Report, June 2014, https://nrf.com/sites/default/files/Port%20Closure%20Full%20Report.pdf

Chapter 7

1 Portions of this chapter are based on a published article: A. Hecht and J. Fiksel, "Environment and Security," *Encyclopedia of Earth*, September 2011, www.eoearth.org/view/article/51cbf17c7896bb431f6a5af0

2 US Department of Defense Newsletter, http://www.defense.gov/news/newsarticle.aspx?id=123398

3 D. H. Meadows, J. Randers, D.L. Meadows, and W. W. Behrens III, *The Limits to Growth*, Universe Books, 1974.

4 UN General Assembly, *Report of the World Commission on Environment and Development: Our Common Future*, Oxford Press, 1987, p. 19.

5 National Security Strategy of the United States, 1991, http://fas.org/man/docs/918015-nss.htm

6 National Security Strategy of the United States, 2002, http://www.comw.org/qdr/fulltext/nss2002.pdf, p. 21.

7 US EPA Science Advisory Board, *Beyond the Horizon: Using Foresight to Protect the Environmental Future*, EPA-SAB-EC-95-007, US EPA, January 1995, p. 29.

8 J. Rockström, W. Steffen, K. Noone, Å. Persson, F. Chapin, E. Lambin, T. Lenton, M. Scheffer, C. Folke, H. Schellnhuber, et al., "A Safe Operating Space for Humanity," *Nature* 461 (2009): 472–475.

9 R. M. Hassan, R. Scholes, and N. Ash, *Ecosystems and Human Well-Being: Current State and Trends: Findings of the Condition and Trends Working Group*, Island Press, December 14, 2005.

10 CNA Corporation, *National Security and the Threat of Climate Change*, CNA, 2007, http://www.cna.org/sites/default/files/news/FlipBooks/Climate%20Change%20web/flipviewerxpress.html

11 J. Fiksel, *Design for Environment: A Guide to Sustainable Product Development*, 2nd ed. McGraw-Hill, 2009.

12 National Commission on the BP Deepwater Horizon Oil Spill and Offshore Drilling, "The Gulf Oil Disaster and the Future of Offshore Drilling," *Encyclopedia of the Earth*, January 2011, http://www.eoearth.org/view/article/162358/

13 United Nations Department of Economic and Social Affairs, "2014 Revision of the World Urbanization Prospects," http://www.un.org/en/development/desa/publications/2014-revision-world-urbanization-prospects.html

14 United Nations, *The Millennium Development Goals Report*, 2014, http://www.un.org/millenniumgoals/2014%20MDG%20report/MDG%202014%20English%20web.pdf

15   A. D. Hecht, "The Next Level of Environmental Protection," *Sustainable Development Law and Policy* 8 (Fall 2009).

16   A. McCartor and D. Becker, *Blacksmith Institute's World's Worst Pollution Problems Report 2010*, Blacksmith Institute and Green Cross, 2010, http://www.worstpolluted.org/files/FileUpload/files/2010/WWPP-2010-Report-Web.pdf

17   See the principles of green chemistry and engineering promulgated by Yale University, http://greenchemistry.yale.edu/green-chemistry-green-engineering-defined

18   Helpful information was provided by Neil Hawkins and Mark Weick of Dow Chemical.

19   J. L. DiMuro, F. M. Guertin, R. K. Helling, J. L. Perkins, and S. Romer, "A Financial and Environmental Analysis of Constructed Wetlands for Industrial Wastewater Treatment," *Journal of Industrial Ecology* 18, no. 5 (2014): 631–640.

20   Joint Industry White Paper, "The Case for Green Infrastructure," June 2013, http://www.nature.org/about-us/the-case-for-green-infrastructure.pdf

Chapter 8

1   E. Beinhocker, "The Adaptable Corporation," *McKinsey Quarterly* 2 (2006).

2   L. Välikangas, *The Resilient Organization: How Adaptive Cultures Thrive Even When Strategy Fails*, McGraw-Hill, 2010, p. 110.

3   G. Hamel and L. Välikangas, "The Quest for Resilience," *Harvard Business Review* 81, no. 9 (2003): 54.

4   K. Weick and K. Sutcliffe, *Managing the Unexpected: Resilient Performance in an Age of Uncertainty*, Jossey-Bass, 2007.

5   S. L. Brown and K. M. Eisenhardt, *Competing on the Edge: Strategy as Structured Chaos*, Harvard Business School Press, 1998.

6   L. Välikangas, *The Resilient Organization*, p. 92.

7   P. H. Longstaff, "Is the Blame Game Making Us Less Resilient?," *Proceedings of the First International Symposium on Societal Resilience*, Homeland Security Studies and Analysis Institute, 2011.

8   Resilient Organizations, http://www.resorgs.org.nz

9   C. A. Lengnick-Hall, T. E. Beck, and M. L. Lengnick-Hall, "Developing a Capacity for Organizational Resilience through Strategic Human Resource Management," *Human Resource Management Review* 21 (2011): 243–255.

10   E. Hollnagel, J. Pariès, D. Woods, and J. Wreathall, eds., *Resilience Engineering in Practice: A Guidebook*, Ashgate, 2010.

11   Helpful information was provided by the following individuals at AEP: Doug Buck, Sylvette Gilbert, Phil Lewis, Jim Nowak, Craig Rhoades, and Laura Thomas. The case study was coauthored by Iliana Filyanova of Ohio State University's Fisher College of Business.

Chapter 9

1   Y. Berra, "The Yogi Book: I Really Didn't Say Everything I Said!," Workman Publishing, 1998, pp. 118–119.

2   P. Sydelko, S. Ronis, and L. Guzowski. "Energy Security as a Wicked Problem—A Foresight Approach to Developing a Grand Strategy for Resilience," *Solutions*, 5, no. 5 (September/October 2014): 11–16.

3   http://www.shell.com/global/future-energy/scenarios/new-lens-scenarios.html

4   J. Fiksel, *Design for Environment: A Guide to Sustainable Product Development*, 2nd ed., McGraw-Hill, 2009.

5   T. J. Pettit, K. L. Croxton, and J. Fiksel, "Ensuring Supply Chain Resilience: Development and Implementation of an Assessment Tool," *Journal of Business Logistics* 34, no. 1 (March 2013): 46–76. Performance volatility was measured by the coefficient of variation (i.e., the ratio of the standard deviation to the mean).

6   B. Walker, C. S. Holling, S. R. Carpenter, and A. Kinzig, "Resilience, Adaptability and Transformability in Social-Ecological Systems," *Ecology and Society* 9 (2004), http://www.ecologyandsociety.org/vol9/iss2/art5/

7   J. McDaniel and J. Fiksel, "The Lean and Green Supply Chain: A Practical Guide for Materials Managers and Supply Chain Managers to Reduce Costs and Improve Environmental Performance," US EPA Office of Pollution Prevention and Toxics, EPA 742-R-00-001, January 2000.

8   K. J. Arrow, P. Dasgupta, and K.-G. Maler, "Evaluating Projects and Assessing Sustainable Development in Imperfect Economics," *Environmental and Resource Economics* 26, no. 4 (2003): 647–685.

9   P. Kumar, ed., *The Economics of Ecosystems and Biodiversity: Ecological and Economic Foundations*, Earthscan, 2010.

10   J. Fiksel, E. Irwin, and M. Gnagey, "Resilience Economics: A Systems Approach," *Proceedings of the NIST-ASCE Workshop on Economics of Community Disaster Resilience*, National Institute of Standards and Technology, forthcoming.

11   Helpful information was provided by Lou Ferretti at IBM.

12   C. Krosinsky, "How Going Green Can Make You Rich," *Daily Beast*, October 16, 2011, http://www.erb.umich.edu/News-and-Events/news-events-docs/11-12/going-green-can-make-you-rich.pdf

Chapter 10

1   T. Brown, *Change by Design: How Design Thinking Transforms Organizations and Inspires Innovation*. HarperCollins, 2009, p. 229.

2   C. M. Christensen, *The Innovator's Dilemma: When New Technologies Cause Great Firms to Fail*, Harvard Business School Press, 1997.

3   D. Rainey, *Product Innovation: Leading Change through Integrated Product Development*, Cambridge University Press, 2005.

4   J. Fiksel, *Design for Environment: A Guide to Sustainable Product Development*, 2nd ed., McGraw-Hill, 2009.

5   J. Fiksel, "Designing Resilient, Sustainable Systems," *Environmental Science and Technology*, December 2003.

6   M. Porter and M. Kramer, "Strategy and Society," *Harvard Business Review*, December 2006.

7   A. Lovins and H. Lovins, *Brittle Power*, Brick House, 1984.

8   J. E. Lovelock, *The Ages of Gaia: A Biography of our Living Earth*, Norton, 1988.

9   C. K. Prahalad and S. L. Hart, "The Fortune at the Bottom of the Pyramid," *strategy+business* 26 (2002).

10   US Resilience Project, http://usresilienceproject.org/best-practices/supply-chain-resilience/

11   E. D. Williams, R. U. Ayres, and M. Heller, "The 1.7 Kilogram Microchip: Energy and Material Use in the Production of Semiconductor Devices," *Environmental Science and Technology* 36 (2002): 5504.

12   G. J. Persley and J. N. Siedow, *Applications of Biotechnology to Crops: Benefits and Risks*, Council for Agricultural Science and Technology, 1999.

Chapter 11

1   P. Senge, B. Smith, N. Kruschwitz, J. Laur, and S. Schley, *The Necessary Revolution: How Individuals and Organizations Are Working Together to Create a Sustainable World*, US Green Building Council, 2008, p. 6.

2   D. N. Bresch, J. Berghuijs, R. Egloff, and R. Kupers, "A Resilience Lens for Enterprise Risk Management," in *Turbulence: A Corporate Perspective on Collaborating for Resilience*, ed. R. Kupers, Amsterdam University Press, 2014.

3   National Research Council, *Pathways to Urban Sustainability: Research and Development on Urban Systems*, National Academies Press, 2010.

4   President's State, Local, and Tribal Leaders Task Force on Climate Preparedness and Resilience: Recommendations to the President, November 2014, http://www.white house.gov/administration/eop/ceq/initiatives/resilience/taskforce

5   L. S. Fuerth and E. M. H. Faber, "Anticipatory Governance: Practical Upgrades," National Defense University and George Washington University, 2012.

6   Committee on Increasing National Resilience to Hazards and Disasters and Committee on Science, Engineering, and Public Policy, National Academies, *Disaster Resilience: A National Imperative*, National Academies Press, 2012.

7   P. C. Evans and P. Fox-Penner, "Resilient and Sustainable Infrastructure for Urban Energy Systems," *Solutions*, Fall 2014.

8   New York State 2100 Commission, "Recommendations to Improve the Strength and Resilience of the Empire State's Infrastructure," 2013, http://www.governor.ny.gov/sites/governor.ny.gov/files/archive/assets/documents/NYS2100.pdf

9   See http://army-energy.hqda.pentagon.mil/programs/netzero.asp.

10   Helpful information was provided by Peter Williams at IBM.

11   http://www.unisdr.org/2014/campaign-cities/Scorecard%20FAQs%20March%2010th%202014.pdf

12   http://www.unisdr.org/campaign/resilientcities/assets/documents/ten-essentials.pdf

Chapter 12

1   B. Gates, *Business @ the Speed of Thought: Succeeding in the Digital Economy*, Warner Books, 1999, p. 131.

2   Sustainable development is commonly defined as "development that meets the needs of the present without compromising the ability of future generations to meet their own needs." World Commission on Environment and Development, *Our Common Future*, Oxford University Press, 1987, p. 43.

3   Millennium Ecosystem Assessment, Synthesis Report, *Ecosystems and Human Well-Being*, Island Press, 2003.

4   Global Footprint Network, "China Ecological Footprint Report 2012: Consumption, Production and Sustainable Development," http://www.footprintnetwork.org

5   http://www.ceres.org/about-us/our-history/ceres-principles

6   http://www.unglobalcompact.org/aboutthegc/thetenprinciples/index.html

7   http://thenaturalstep.org/the-system-conditions

8   The triple bottom line consists of environmental, social, and economic dimensions of corporate performance. J. Elkington, *Cannibals with Forks: The Triple Bottom Line of 21st Century Business*, Capstone Publishing, 1997.

9   J. Fiksel, "Revealing the Value of Sustainable Development," *Corporate Strategy Today* 7/8 (2003).

10   Some researchers consider environmental impacts to be "acceptable" if resources are not consumed faster than the rate of replenishment and if waste generation does not

exceed the carrying capacity of the surrounding ecosystem. K. H. Robèrt, *The Natural Step: A Framework for Achieving Sustainability in Our Organizations*, Pegasus, 1997.

11 B. Bakshi and J. Fiksel, "The Quest for Sustainability: Challenges for Process Systems Engineering," *AIChE Journal*, June 2003, p. 1350.

12 World Business Council for Sustainable Development, *Mobility 2030: Meeting the Challenge of Sustainability*, Geneva, 2003, p. 5.

13 H. Daly, *Steady-State Economics*, 2nd ed., Island Press, 1991.

14 UNEP Global Environmental Outlook 4, http://www.unep.org/geo/geo4.asp

15 A. Lovins, *Reinventing Fire*, Rocky Mountain Institute, 2011.

16 R. Dobbs, J. Oppenheim, and F. Thompson, "Mobilizing for a Resource Revolution," *McKinsey Quarterly*, January 2012.

17 A. D. Hecht, J. Fiksel, S. C. Fulton, T. F. Yosie, N. C. Hawkins, H. Leuenberger, J. Golden, and T. E. Lovejoy, "Creating the Future We Want," *Sustainability: Science, Practice, and Policy* 8, no. 2 (Summer 2012): 63.

18 A. Winston, "Resilience in a Hotter World." *Harvard Business Review*, April 2014.

19 C. Cutler, "Energy Return on Investment (EROI)," *Encyclopedia of Earth*, 2011.

20 P. Fox-Penner, *Smart Power: Climate Change, the Smart Grid, and the Future of Electric Utilities*, Island Press, 2010.

21 United Nations Environment Programme, *Melbourne Principles for Sustainable Cities*, 2002.

22 W. McDonough and M. Braungart, *Cradle to Cradle: Remaking the Way We Make Things*, North Point Press, 2002.

Index

Page numbers followed by "f" and "t" indicate figures and tables.

DATE DUE

SYLVIA PORTER'S 385 TAX-SAVING TIPS

HOW TO PROFIT FROM THE NEW TAX LAWS

Other Avon Books by
Sylvia Porter

SYLVIA PORTER'S LOVE AND MONEY
SYLVIA PORTER'S NEW MONEY BOOK FOR THE 80'S
SYLVIA PORTER'S YOUR OWN MONEY